Novel Aspects of Cryopreservation

Novel Aspects of Cryopreservation

Edited by **Marianne Wilde**

New York

Published by Callisto Reference,
106 Park Avenue, Suite 200,
New York, NY 10016, USA
www.callistoreference.com

Novel Aspects of Cryopreservation
Edited by Marianne Wilde

© 2015 Callisto Reference

International Standard Book Number: 978-1-63239-480-4 (Hardback)

Contents

Preface

I am honored to present to you this unique book which encompasses the most up-to-date data in the field. I was extremely pleased to get this opportunity of editing the work of experts from across the globe. I have also written papers in this field and researched the various aspects revolving around the progress of the discipline. I have tried to unify my knowledge along with that of stalwarts from every corner of the world, to produce a text which not only benefits the readers but also facilitates the growth of the field.

This book presents significant aspects of cryopreservation. Lately, there have been certain severe tectonic shifts in cryobiology though not visible on the surface but will have significant impact on both the advancement of novel cryopreservation techniques and the future of cryobiology. This comprehensive book discusses the existing applications and emerging practical protocols for the purpose of cryopreservation along with the description of the novel cryobiological ideas. The topics have been loosely divided into three sections namely, Stem Cells and Cryopreservation in Regenerative Medicine, Human Assisted Reproduction Techniques (ART), and Farm/Pet/Laboratory Animal ART.

Finally, I would like to thank all the contributing authors for their valuable time and contributions. This book would not have been possible without their efforts. I would also like to thank my friends and family for their constant support.

Editor

Part 1

Stem Cells and Cryopreservation in Regenerative Medicine

Cryopreserved Musculoskeletal Tissue Bank in Dentistry: State of the Art and Perspectives

Luiz Augusto U. Santos[1], Alberto T. Croci[2], Nilson Roberto Armentano[3],
Zeffer Gueno de Oliveira[4], Arlete M.M. Giovani[5],
Ana Cristina Ferreira Bassit[6], Graziela Guidoni Maragni[7],
Thais Queiróz Santolin[7] and Lucas da Silva C. Pereira[8]

1. Introduction

Maxillary and mandibular bone loss has long been a challenge to dental surgeons who seek to reconstruct these lost segments. These lesions lead to deformation of some maxillary and mandibular areas which interferes in the functional rehabilitation process of these structures. The most common cause of these lesions is prolonged use of total prostheses in a large part of the Brazilian population and the searches for surgical techniques and bone substitutes are today proposed and studied by the academic class. In this context, Brazil is starting to distribute allogeneic tissue obtained, processed and qualified by musculoskeletal tissue banks. Such banks already have experience in dispensing tissue to the orthopedic area, which has been using reconstructive techniques with allografts for many years. The first studies proposing the use of bone substitutes for replacement of these faulty parts commenced in the decades subsequent to 1860. (Carrel, 1912;Groves, 1917; Sharrard, Collins, 1961; Urist, 1965; Fischer, 1998; Tomford, 2000).

After the verification of the disadvantages in the use of autologous tissues for this purpose, such as the increase in donor morbidity, greater risk of nerve lesion and of infection inherent to the second surgical procedure and limitation in the availability of the tissue in quantity and variety, the use of homologous tissue became another option that was gradually indicated (Cunningham, Reddi, 1992; Tomford, 2000).

[1]Institute of Orthopedics and Traumatology, Hospital das Clínicas of the School of Medicine of the University of Sao Paulo, dentist, Tissue Bank Technical Responsible and. Sao Paulo/SP, Brazil
[2]Institute of Orthopedics and Traumatology, Hospital das Clínicas of the University of São Paulo school of Medicine, Professor and Tissue Bank Director - Sao Paulo/SP, Brazil
[3]School of Dentistry of the University of Santo Amaro- São Paulo, Brazil
[4]Orthopedic Nurse Specialist. São Paulo/SP, Brazil
[5]Nurse, Institute of Orthopedics and Traumatology, Hospital das Clínicas of the University of São Paulo school of Medicine, Tissue Bank Coordinator - São Paulo/SP, Brazil
[6]Veterinarian, Tissue Bank Researcher, University of Florida, Gainesville, FL – Flórida- US
[7]Nurse, Institute of Orthopedics and Traumatology, Hospital das Clínicas of the University of São Paulo school of Medicine, Tissue Bank Team - São Paulo/SP, Brazil
[8]Dental Student, Institute of Orthopedics and Traumatology, Hospital das Clínicas of the University of São Paulo school of Medicine, Tissue Bank Team - São Paulo/SP, Brazil

The good results with the clinical application of allografts in dentistry motivate their use on an increasing scale, until in 2005 Dentistry came into the scene with the use of tissues in maxillary and mandibular pre-prosthetic surgery. A consensus between the National Transplant System and the Federal Board of Dentistry allows the use of allografts by specialists in the areas of Implant Dentistry, Periodontics and Oral and Maxillofacial. The tissue banks, in turn, prepare a tissue processing line geared toward dental needs with a focus on quality control and traceability.

Thus usage has become both abrupt and a tendency in the last 5 years (RBT, 2006-2010). In spite of a significant number of bone transplants in the dental area with good clinical results, the dental profession is still lacking information about activities that involve the area of tissue banks, particularly in the rigid quality control and traceability. Such activities are founded on international standards[2], literature[3] and legislation[4] and implemented according to Good Manufacturing Practice- GMP.

We consider it very important to gather epidemiological data on bone transplants in dentistry, elucidating the size and the limits of this type of treatment that is already considered a tendency in our field. In addition, to report on our perspectives of investigation into the efficacy and safety of the use of allografts, with tests that can enable us to expand our knowledge about the osseointegration of allografts. In other words, knowledge that allows us to reach what we consider most important in dental treatments: the predictability of treatment.

2. Bone tissue

Bone tissue is composed of two portions: 1. Organic, consisting of intrinsic bone cells (osteoblasts, osteoclasts and osteocytes) and the organic matrix synthesized thereby; 2. Inorganic, consisting of hydroxyapatite, deposited amorphously in an initial phase and that in a short space of time is converted into another crystalline hydroxyapatite. Organic matrix corresponds to 35% of the bone volume and inorganic matrix to 65%.

In spite of the resistance and hardness, bone tissue is very plastic and has a high capacity to remodel through various situations to which it is submitted, such as fractures, lesions and bone loss. The bone tissue regeneration process starts from important biological reactions, triggered by the actual tissue lesion. Grafting triggers a mechanism of migration of the bone cells belonging to the receptor bed to the inside of the graft, with the purpose of resorbing it and replacing it with neoformed bone.

---

[2]European Association of Tissue Banks. Common Standards for Tissues and Cells Banking: Berlin: European Association of Tissue Banks; 2004.
American Association of Tissue Banks. Standards for Tissue Banking. 11th ed. McLean : American Association of Tissue Banks; 2007
[3]Phillips GO, Strong DM, Versen RV, Nather A. Advances in Tissue Banking. Vol. 4. World Scientific . New Jersey, 2000.
Bancroft JD, Stevens A. Theory and practice of histological techniques. Fourth Edition. Churchill Livingstone. United Kingdom, 1999.
[4]Law n.9434 of February 5, 1997; Decree n.2268 of June 30, 1997;Administrative Ruling n.1686 of September 20, 2002; Resolution n. 220 of December 27, 2006; Administrative Ruling n. 2600 of October 21, 2009.

The cells belonging to the bone tissue are the osteoblasts, osteocytes and osteoclasts.

The osteoblasts are cuboid, elongated cells of mesenchymal origin that are located in the bone margins; their function is to produce the organic matrix of the bone tissue. In reduced activity these cells assume a more slender shape. The osteocytes are encapsulated osteoblasts, which after maturation became imprisoned inside the mineralized matrix, but that still maintain contact with other cells through cytoplasmic ramifications, thus maintaining physiologic functionality of the tissue (Junqueira, Carneiro, 1999; Davies 2000). This contact with surface cells such as the osteoblasts and lining cells is related bone structure maintenance and to the physiological responses that lead to tissue formation or resorption (Aubin et al., 2006).

The osteoclasts are giant cells with multiple nuclei and their function is related to resorption. In synergy with the osteoblasts they promote bone remodeling.

The interaction and the synergism among bone cells is called creeping substitution, and this occurs through three essential cellular events: osteogenesis (cellular event that favors the synthesis of bone matrix by the osteoblasts), osteoinduction (ability to induce the migration of mesenchymal cells and their differentiation into osteoblasts) and osteoconduction (ability of the tissue to serve as a mold or guide for the cellular processes involved in bone tissue repair).

Moreover, as is the case with others, the bone cells pass through the stages of the cell cycle, which range from formation to cell division (mitosis). Mitosis is susceptible to external interferences, and the cell can either enter a state of rest or continue to split cyclically (Urist, 1965; Enneking et al., 1975; Junqueira, Carneiro, 1999; Perren, Claes, 2002).

3. Bone transplantation

The term transplant is not widely used by the dental community to refer to the use of bone tissue. The common term is bone grafting. The bone graft can receive a nomenclature and be classified according to the origin of its obtainment and on the implant site (Table 1).

Autologous graft or autograft	Graft of tissue from one site to another in the same individual
Isograft	Graft between people with the same genotype (homozygotes; e.g. identical twins)
Allograft or homologous graft	Graft between individuals of the same species with disparate genotype
Xenograft or heterologous graft	Graft from one individual of a species onto a different species

Table 1. Classification of grafts according to their nature. Source: Drumond, 2000

4. Clinical application of grafts over the years

The use of bone tissue for replacement of bone losses is not a recent procedure. Since last century there have been accounts of the use of these tissues in humans and in experiments with animals as a means of assessing their efficacy.

In these studies, many treatments with the use of bone grafts of autologous and homologous nature have been proposed over the years. The first homologous bone transplant is described by William MacEwen in 1878 (Giovani, 2005). At this time the treatment of osteomyelitis was performed by means of surgical resections of the infected segments. Homologous tibial segments (obtained from patients submitted to osteotomies) were used in the reconstruction of a bone defect caused by the resection of part of the humerus of a young man suffering from osteomyelitis (Tomford, 2000). Encouraging results of the consolidation of the bone graft with the receptor bed (MacEwen, 1909) motivated researchers back then. In this period, there was no consensus about which bones can specifically be used for transplantation. Tissues were obtained randomly from donors that were victims of fractures, resections and amputations. For the storage and processing of these tissues, professionals used protocols and equipment that today are not the best suited to this purpose (Tomford, Mankin, 1999; Tomford, 2000). Nowadays a large portion of these studies has only historical value with respect to the pioneer spirit of these researchers.

The clinical use of allografts during this period was in low demand, and occurred on an experimental basis at some centers from all around the world. With the availability of antibiotic therapy, changes occurred in the indication of these tissues, and patients with osteomyelitis were then submitted to pharmacological treatments, to the detriment of surgical procedures (Tomford, 2000). Thus surgical resections of infected segments cease to be a priority, and allografts are then used in the reconstructions of bone defects caused by tumor resections. This change causes the studies from the time to evolve as well, and to give more detailed accounts of the cellular processes involved in the osseointegration of grafts. The consensuses of these first studies serve as guidelines for the first grafts performed at that time.

The use of cryopreserved allografts presents some advantages over autologous tissue, such as the availability of the necessary quantity of tissue and a decrease in postoperative morbidity. As regards morphology, there are some differences in the vascularization process of the cortical and spongy bone grafts. In the cortical bone, the repair is started by the action of osteoclasts and in the spongy bone, by the osteoblasts. Another difference lies in the revascularization time, which is slower for cortical bone and faster for spongy bone.

At that time, Urist (1965) was already describing that the osteoprogenitor cells responsible for the bone repair process are derived from monocytes, present in an elevated number in the repair zone coming from the bone marrow, and that the osseointegration of grafts is achieved, since primitive cells not yet differentiated can differentiate into viable osteoblasts from osteoinductive substances. Perivascular mesenchymal cells disaggregate and migrate to the grafting area, where they reaggregate, proliferate and differentiate to form new bone. (Urist et al, 1983)

Some substances secreted by certain cell types interfere in or even modulate the cellular processes of osseointegration, today known as Cytokines or growth factors.

Urist made important discoveries in this area back in 1965, even proposing a new bone processing procedure aimed at the removal of a calcified layer (demineralization) from the matrix, making this graft more osteoinductive. At the time this kind of tissue was called Demineralized Bone Matrix (DBM), and the name is still in use today. Later, it was understood that part of this capacity is the responsibility of the superfamilies of proteins (TGF), including the morphogenetic protein (BMP) present in the bone matrix. (Malafaya et al, 2002).

The decade of 1980 was marked by the advent of the acquired immunodeficiency syndrome (AIDS) that gives rise to discussions on the safety of the clinical use of homologous tissues. The biological risk of disease transmission between tissue donors and recipients is the topic of greatest relevance and importance in the period. The tissue banks existing at that time were encouraged to review donor selection protocols with the objective of avoiding the transmission of these diseases. This encouragement was provided mainly by the international public health regulatory agencies, such as the FDA (Food and Drug Administration) and other institutions related to the Haemovigilance and tissue transplantation systems. The result was a standardization of the internal processes of banks with respective preparation of standards by the main global tissue bank associations (American Associating of Tissue Banks - AATB and European Association of Tissue Banks-EATB), which contributes to the gain of quality of tissue made available by these services (Nather, 1991; Galea, Kearney, 2005; Santos, 2007).

With the availability of more reliable grafts by the tissue banks, their use, biological behavior and indication by surgeons become a viable treatment option. (Santos, 2011).

Similar to the repair process in fractures and in the development of the musculoskeletal system, the osseointegration of grafts takes place after a selection of primordial cells that are differentiated into osteoblasts under the influence of osteogenic factors. (Thies et al., 1992). The main objective expected in the use of grafts is the ability to selectively induce the primordial events of the integration process, such as osteoinduction, osteoconduction and osteogenesis (Lindhe et al., 1997).

According to Tomford and Mankin (1999), for cortical bone grafts to be incorporated into the receptor bed, there must be revascularization of this bed. When this process does not occur, the repair area loses balance in resorption and the graft might suffer fatigue fractures. Spongy grafts are consolidated more quickly.

Boldt et al. (2001) evaluate the use of frozen bone graft in 173 acetabular reconstructions and 79 femoral reconstructions in humans. Femoral heads obtained from a local bone bank are particulated and impacted in the faults. After a mean follow-up period of four years, they report acetabular clinical stability in 97.2 % of the cases, graft incorporation in 74% in the acetabula and 61% in the femurs, according to radiological analysis. They conclude that the results obtained with the use of impacted grafts are promising, except for the reconstruction of type III acetabular defects, where a reinforcement cage is recommended.

Janssen et al. (2001) studied the use of homologous cortical rings obtained from femoral diaphysis for reconstruction of intervertebral discs of the lumbar spine in 137 patients. The results show that arthrodesis was achieved in 94% of the cases and they do not report signs of resorption.

Weyts et al. (2003) analyze samples of femoral heads collected after the primary arthroplasty of two human donors. The tissues are cryopreserved at - 80 °C for a minimum period of six months, according to the protocols of the American and European Tissue Bank Associations. After this period, these tissues are biopsied and submitted to cell culture (for survival observation) and PCR (polymerase chain reaction) for genetic screening. The results show the presence of live cells belonging to the donors in the analyzed samples.

Lavernia et al. (2004) researched the adoption and use of allografts by orthopedic surgeons in 340 U.S. hospitals. The frozen graft from tissue banks certified by the American Association of Tissue Banks - AATB is used the most often by orthopedists for the treatment of knee and hip bone loss.

Schreurs et al. (2005) conduct a study to evaluate the use of homologous bone graft from a tissue bank in the reconstruction of 33 femoral defects. Bone graft impaction precedes the fixation of the femoral nail. After a minimum follow-up of eight years there is functional improvement of the joint according to the Harris hip score (from 49 to 85 points from pre- to postoperative period) and good survival according to the Kaplan-Meier method. Although four patients have had femoral fractures, the authors conclude that the graft impaction technique and use of cemented femoral nail results in excellent survival for eight to thirteen years.

Cabrita (2007) studies the treatment of infected hip arthroplasties with and without the use of the antibiotic-impregnated cement spacer. For the reconstruction of bone stock, Cabrita uses the massive particulated homologous graft in 60.9% of the patients treated and does not report complications related to their use.

In a review of concepts, Giannoudis et al. (2005) emphasize the advantages of the use of bone grafts in the area of orthopedics and traumatology. They describe the cellular events present in literature that involves the osseointegration process of autologous, homologous grafts and of biocompatible synthetic substitutes. They stress the osteoinductive characteristic of fresh grafts and the osteoconductive characteristic of frozen and lyophilized grafts. Heyligers and Kleim (2005) verify the presence of live cells with growth potential in samples of femoral heads cryopreserved at - 80 °C, over a minimum period of six months. The authors stress the importance of discussing the osteoconductive potential of grafts and highlight the need to investigate the role of these surviving cells from the frozen tissue in the bone formation process after its implantation. Besides the bone cells, other cells and inflammatory factors play a vital role in the bone repair process. The macrophages and substances such as interleukin one, six, eleven (IL-1, IL-6, IL-11), RANKL and osteoprotegerin (OPG) are found during the first three days after the lesion (Gerstenfeld, Einhorn, 2006). As far as macrophages are concerned, Knighton et al. (1982) explain that these cells are present in the repair zone and are capable of producing growth factors, which, in turn, stimulate the neovascularization, proliferation and migration of other cell types, such as the fibroblasts.

5. Evolution of musculoskeletal tissue banks in Brazil

Considering the satisfactory results in the use of allografts obtained from multiple donors of organs and tissues with brain death besides studies showing revascularization, osteointegration and bone formation at sites of graft (Barros Filho, et al., 1989; Croci et al., 2003, Zhang et al., 2004; Dallari et al., 2006, Bitar et al, 2010), an increasing number of orthopedic surgeons and dentists currently opt to use of homologous grafts in our country. This fact is corroborated by the disadvantages already known in the use of autologous tissues, such as the increase in postoperative morbidity, greater risk of infection inherent to the second surgical procedure required for their obtainment, risk of nerve lesion and limitation of the quantity and variety of graft obtained (Smith et al. 1984; Cunningham, Reddi, 1992; Drumond, 2000).

This tendency, combined with the growing number of patients with bone loss that seek specialized orthopedic and dental services, leverages the creation of some Tissue Banks in the country in an experimental manner.

In Brazil, the standardization of Musculoskeletal Tissue Banks is linked to specific laws of our country. We have a General Coordination Office for the National Transplant System – SNT of the Ministry of Health that discusses and prepares, together with technical chambers, legislations involving the use of tissue by the medical and dental community. The guidelines are baséd on protocols already developed by some centers and on the standards of international associations (European Association of Tissue Banks - EATB and American Association of Tissue Banks-AATB). The creation of the first reference centers in the large-scale capture, processing and distribution of musculoskeletal tissue and some of these experiences are described. (Amatuzzi et al., 2000, Amatuzzi et al., 2004). Tissue Banks in Brazil are controlled by the General Management of Blood, other Tissues, Cells and Organs - GGSTO of the National Health Surveillance Agency - ANVISA, an institution similar to the U.S. FDA. This agency focuses its activities on health surveillance and on quality control, traceability, appraisal of risks and of adverse effects involving tissue transplants in the country and its guidelines are published in the form of legislation[5] that also defines Musculoskeletal Tissue Bank as "the service that, with physical facilities, equipment, human resources and adequate techniques, has as its duties the performance of clinical, laboratory and serological triage of tissue donors, the removal, identification, transportation, processing, storage and delivery of bones, soft tissues (cartilage, fasciae, serous membranes, muscle tissue, ligaments and tendons) and their derivatives, of human origin for therapeutic purposes, research and teaching".

6. Activities of a musculoskeletal tissue bank

The description of activities of a musculoskeletal tissue bank is summarized in the algorithm below (**Illustration 01**).

Every activity related to the bank should also be based on the ethical principles inherent to the activities of any organ transplantation. They are:

- **Autonomy and self-determination:** The recipient of tissue from the musculoskeletal system should be provided with information in accessible language about the entire tissue obtainment process, the risks and the chances of success or failure of the treatment. The following stage is the patient's decision, after their evaluation of the information received, set out in an informed consent form.
  Professionals with specific training provide all the required information, using language exempt from complex or technical terms, enabling the patient to achieve easy understand in order to make the final decision. For this document to be authentic, the consent must be free, that is, not caused by coercion. The professionals of a musculoskeletal tissue bank should be objective and impartial while providing guidance to recipients. Every process is recorded in the Recipient's Form.

---

[5]Administrative Ruling no. 211, of March 24, 2003

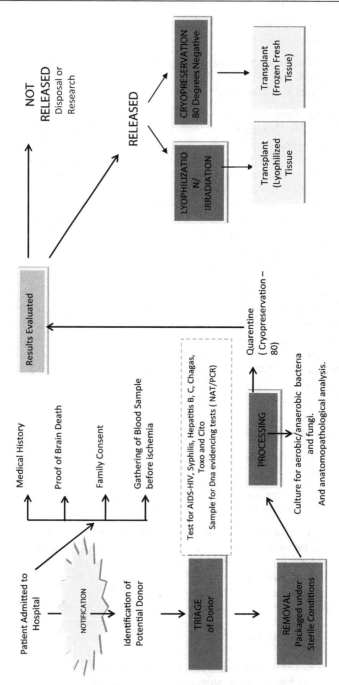

Illustration 01. Algorithm of the Musculoskeletal Tissue Donation and Transplantation Process

- **Justice** – The principle of equal opportunities for the use of available tissues. The definition of ethical parameters in distribution is imperative.
- **Symbolism of the body** – This principle is employed mainly in the reconstruction of the deceased donor's body after removal, which should be performed carefully, with the apparent anatomical parameters respected, thus ensuring that the family receives a body in adequate conditions.

7. Obtaining musculoskeletal tissues

The main source of musculoskeletal tissues is the notification of deceased donors to the Transplant Centers, Organ Service Services, and Hospital Transplant Departments. The teams that receive the organs and tissues are only notified after a series of procedures and exams has been carried out to ascertain brain death and obtain the family's consent for the process of organ and tissue donation. Brain death is initially verified by a neurologist, using techniques of physical and imaging (doppler) exams, which are repeated after six hours in the presence of a family member of the potential donor. Once there is no doubt as to the irreversible diagnosis of brain death, the family members are asked whether they would consider donating their loved one's organs. The family interview is done by trained members belonging to an intra-hospital committee, or by an organization that looks for organs. The entire donation process should be recorded and legally signed before the teams are notified to remove each organ (heart, liver, kidney, pancreas, lung, intestine) and tissues (osteochondral and fascial-ligamentous, skin, vessels, cornea, heart valves). Each team should have clearly-defined criteria for selecting, and at the time of notification, accepting or refusing the donor in question.

For donors of musculoskeletal tissue, the selection follows a rigorous control process, with serological tests for antigens and HIV antibodies, Hepatitis A, B and C, HTLV-1 and 2, Syphilis, Chagas disease, Toxoplasmosis and Cytomegalovirus, as well as state-of-the-art tests for evidenciation of D (Nucleic Acid Amplification – NAT) HIV and Hepatitis B and C, bone marrow aspirate smear of the sternum and iliac crest sample, both for histopathological investigation.

Donors with the following criteria were excluded: orthopedic pathologies, such as osteoporosis, osteonecrosis, rheumatoid arthritis, lupus erythematosus, neoplasias, age group that compromises that characteristics of the tissues, blood transfusions, tattoos or piercings within the period of the immunological window, use of illegal drugs, travel to endemic zones, generalized or localized infections, fractures, open sores on the limbs from which the musculoskeletal tissues are to be removed, or any other situation that places in doubt the quality of these tissues, pursuant to the Brazilian legislation.

The whole procedure is carried out in totally antiseptic conditions, just as in surgery. Special gowns made from synthetic material are used, and all the surgical stages of antisepsis are followed.

The removed tissues are immediately packaged in triple packaging, hermetically sealed, and delivered, under refrigeration (-4°C) to the Tissue Bank.

A very important stage of the capture process is donor reconstruction. The body should be delivered to the family free from any deformation and as close as possible to its appearance before the tissue removal. This is because the fear of deformation has been one of the main causes of refusals of bone donation by the family. For a perfect reconstruction the professionals use prostheses especially developed for this purpose, plaster, sutured and gauze. (**Illustration 02**). This rigorous reconstruction is the most laborious stage in the removal procedure. Areas possibly visible during the funeral (face, anterior side of arm, anterior side of shoulder, etc.) are not approached. All the anatomical parameters are respected, and therefore donor deformation does not occur.

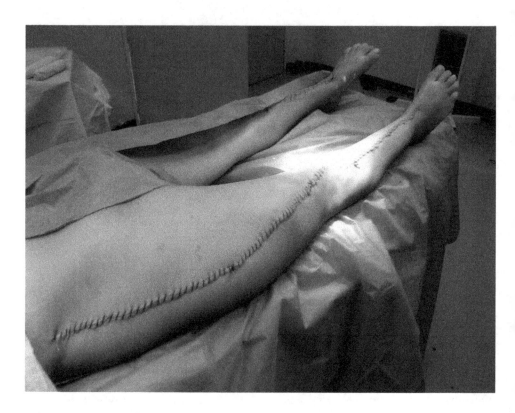

Illustration 02.Limbs reconstructed with prostheses after tissue removal.

8. Processing the musculoskeletal tissues

Once the tissues and organs have been obtained, they are delivered to the BTME in portable refrigerators, with temperature monitoring throughout the transport process. The processing stage is preceded by a planning of the activities necessary to accomplish it, such as provision of materials and instruments, summoning the processing team, defining the preparation and dimensions, according to the needs of the service (waiting list) and requests for orthopaedic and odontological surgeons. This stage is done in a special, classified operating theater (class 100 or ISO 5) equipped with a laminar flow module. (Illustration 03). A Class 100 room means it has purity of 100 particules per m3 of air. For the purposes of comparison, an operating theater should have 10,000 particles/m3 of air.

Illustration 03. Classified processing room (100 particles/m3 air: Class 100/ISO 5; HEPA Filters 99.9 %)

The room also has a pass-through anti-chamber, and all the environments have rigorous control of air particles and positive pressure, to ensure the quality of the tissues processed in it. Specific gowning of the professional team is also necessary, using only non-fabric clothing (Spunbond - Meltblown – Spunbond - SMS) to prevent the dispersion of particles given off by conventional cotton clothing. (**Illustration 04**)

In addition to non-fabric gowning; the team must also adopt certain behaviors. For example, sudden movements, use of cosmetic products and exposure of the skin should be avoided while in this room. Adequate conduct is ensured through special training, not only for the processing team that actually carries out the procedure, but also for other professionals who enter the environment (e.g. for cleaning and maintenance purposes).

A BTME carries out various types of tissue, for use in orthopedic and odontological surgery, and each procedure requires careful planning. (**Illustrations 05 ,06 and 07**)

For the processing of fresh, frozen tissues, a process called mechanical processing is carried out, i.e. removal of the adventitious tissues such as the blood, periosteum, subcutaneous tissue, muscles, fasciae, and fibrotic tissue.

The fragments are then submitted **chemical processing**, where they are immersed in hydrogen peroxide based emulsifying solutions and alcoholic solutions under ultrasound stirring. (**Illustrations 08 and 09**)

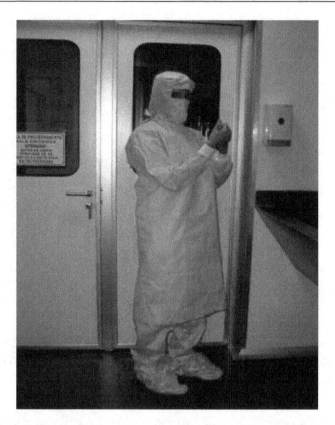

Illustration 04. Specific gowning to work in the classified room.

Illustration 05 and 06. Processing of grafts for use in odontological surgery. The photograph on the left shows modeling of maxillary bone defects based on a resin prototype. The one on the right shows fragments of allografts for use in odontological surgery.

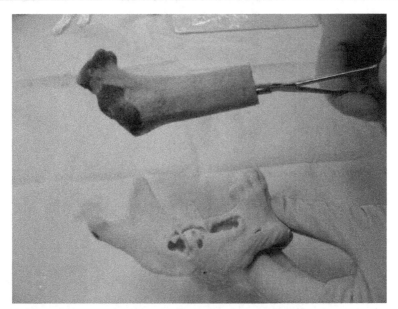

Illustration 07. Modeling of a jaw bone from a segment of proximal femur.

Illustration 08. Ultrasound washing in emulsifying solution.

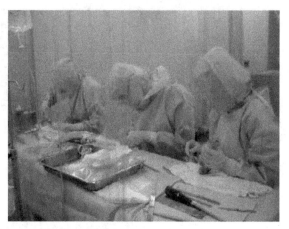

Illustration 09. Processing Team at work.

Immediately afterwards, samples of bone marrow from the long bones and fragments of each tissue submitted to processing are collected from these resulting solutions, and submitted to microbiological processing (general culture, anaerobic and fungal culture) using the direct inoculation technique. Samples are also obtained for histopathological analysis.

Finally, the packaging procedure is begun for all the grafts processed, which are measured (length, height, diameter, weight volume, perimeter), packed in sterile, triple packaging, vacuum sealed, and duly labeled as analysis tissue. (Figure 8). The tissues are labeled with the following information: donor, exams carried out, batch number, item, validity period, type of conservation, and bar code.

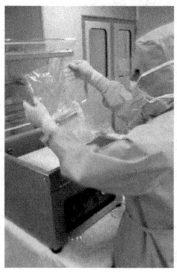

Illustration 10. Vacuum sealing and labeling.

Once they have been labeled, the tissues are submitted to radiography at the BTME, then sent for cryopreservation.

9. Cryopreservation of the musculoskeletal tissues

The musculoskeletal tissue cryopreservation process at - 80°C is described in chapter 1 (VALIDATION OF PRIMARY PACKAGING FOR CRYOPRESERVED MUSCULOSKELETAL TISSUES)

10. Lyophilization of musculoskeletal tissue

The bones can also be processed in their lyophilized form. The lyophilization process should be validated and be, like all the tissue handling procedures, in conformity with the Manual of Good Manufacturing Practice - GMP and in accordance with international standards[6], literature[7] and legislation. The procedure involves the use of an automated lyophilization system composed of a Labconco® **(Illustration 11)** freeze drying, or lyophilization chamber with Condensation Chamber/Vacuum. During lyophilization the tissues remain frozen for the prevention of ice crystal liquefaction inside the matrix. The sublimation process should be validated by analyses of the Residual Moisture by automated thermogravimetric method. The lyophilization process is divided into 2 stages: Primary and Secondary Drying. In the primary phase, the largest fraction of water present in the matrix in its solid state (ice crystals) is removed by sublimation induction and gaseous migration. This induction is achieved by the driving force resulting from the difference of pressure gradient between the lyophilization chamber and condenser. The heat generated by this gaseous transportation should be controlled continually by digital sensors strategically positioned inside the lyophilization chamber.

At the end of the sublimation, the aliquot of unfrozen water linked to the organic components of the matrix (proteins) is then removed in the secondary phase, with an increase of pressure in the lyophilization chamber followed by the gradual increase of temperature at positive levels.

The analysis and control of residual moisture are essential to ensure the integrity of the protein matrix. The residual moisture is determined by thermogravimetric method, using an Ohaus® **(Illustration 12,13)** moisture analyzer. This method analyzes the initial weight of the sample on precision scales, followed by the promotion of heating with continuous recording of evaporation and weight. The percentage of residual moisture **(RM)**, solid mass **(SM)**, initial weight **(IW)** and final weight **(FW)** are analyzed in this method. The limit of RM is < **6** % as described by literature (Phillips, 2000).

---

[6]European Association of Tissue Banks. Common Standards for Tissues and Cells Banking: Berlin: European Association of Tissue Banks; 2004.
American Association of Tissue Banks. Standards for Tissue Banking. 11th ed. McLean : American Association of Tissue Banks; 2007
[7] Phillips GO, Strong DM, Versen RV, Nather A. Advances in Tissue Banking. Vol. 4. World Scientific . New Jersey, 2000.
Bancroft JD, Stevens A. Theory and practice of histological techniques. Fourth Edition. Churchill Livingstone. United Kingdom, 1999.

Illustration 11. Lyophilization chamber: Labconco® equipment from the Tissue Bank of Hospital das Clínicas – Universidade de São Paulo.

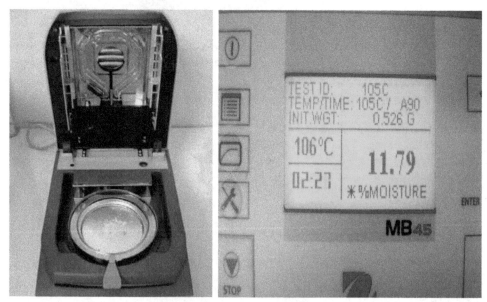

Illustrations 12,13. Residual Moisture Analyzer: Ohaus® equipment belonging to the Tissue Bank of Hospital das Clínicas – Universidade de São Paulo.

The result of the lyophilization is a dry tissue, conservable at room temperature and that should receive final sterilization by radiation. The lyophilized tissue radiation process is carried out in a Multi-purpose Irradiator, with gamma radiation from sources of ^{60}CO. The appropriate radiation dose is 25 kGy, with a dose rate of approximately 7 kGy/h. In order to avoid temperature variation, the samples are radiated in the presence of cooling elements, keeping the temperature between 4 and 8 °C. Red Perspex dosimeters are used for dose control.

Time Interval	Process	Indicators	Sample Temperature
2 hours	Pre-freezing of the lyophilization chamber	Temp. Chamber: – 40 °C	– 80 °C
10 minutes	Vacuum	Chamber Temp.: – 30 °C Chamber Vacuum : 8 x 10 $^{-3}$ mBar Condenser Vacuum : 2 x 10 $^{-3}$ mBar	– 65 °C
6 hours	Primary Drying	Chamber Temp.: – 30 °C Chamber Vacuum: 8 x 10 $^{-3}$ mBar Condenser Vacuum : 1 x 10 $^{-3}$ mBar **Analysis Final Moisture of Phase 01** RM: 6.26% Initial Weight: 0.527g Final Weight: 0.494g Solid: 93.74% **Analysis Temperature: 105°C**	– 40 °C
4 hours	Secondary Drying	Chamber Temp.: up to + 5 °C Chamber Vacuum: 7 x 10 $^{-3}$ mBar Condenser Vacuum: 1 x 10 $^{-3}$ mBar **Analysis Final Humidity of Phase 02** RM: 2.32% Initial Weight: 0.732g Final Weight: 0.715g Solid: 97.68% Analysis Temperature: 105°C	+ 5 °C

Table 1. Primary and Secondary Drying Process

11. Storage, distribution and quality control of tissue

At the end of any processing mode, the procedures should be documented in a specific file. This file is used to contain all the documentation related to the process, including records of the participant team, inputs used, documents evidencing sterility of instruments and other records that are part of the quality control program and that provide subsidies for traceability of the processes. This stage is imperative in the legislation and recommended by the Standards and Quality Control Manuals.

The tissue stock can be kept both frozen and at room temperature (lyophilized) following the same standards used by the global tissue banks and associations. Other processing methods have been investigated with the purpose of reducing the costs related to banking and maintenance. The glycerolization of bone tissue is presented as a processing methodology able to maintain the viability of the matrix and to prevent bacterial growth, besides enabling storage at room temperature (Giovani, 2006).

Until the transplantation occurs, all the processed tissues should be submitted to rigid quality assurance criteria. It is necessary to have an evaluation of all the data pertaining to the donor, results of exams, maintenance and control of equipment, material and instruments used in all the phases of each procedure. Management software can be used to record all the stages, allowing the fast retrieval of information such as the particulars of the donor, lot, shelf life, exams and status of the tissue (analysis, released, excluded, used) making it easier to trace each graft processed and made available, especially in the presence of evidence of an adverse effect and implementation of corrective and preventive actions.

For a lot of grafts under analysis to be released for use, the qualified technical professional from the Tissue Bank must analyze the results of all the exams performed: NAT or PCR serology for HIV, HBC and HCV, General Culture, Anaerobic Culture, Fungal Culture, Anatomopathological Exam and radiology reports. These exam reports are ultimately evaluated and released by the Qualified Clinical Professional of the service.

Besides exams, it is necessary to consider an evaluation of the printed records of temperature during the banking period. The service should have equipment that detects temperature oscillations even at a distance (satellite monitoring system).

Moreover, the installation of buzzers at strategic points of the hospital as well as CO_2 backups ensures the reliability of the system.

After the release of each lot, there should also be a final inspection of each tissue, besides the substitution of tags of tissues under analysis by replaced. The banking logistics of the tissues in the ultra-low temperature (ULT) freezers considers the tissue type and search agility.

Services that execute rigorous quality control use annotation systems featuring checklists with double checking and consent.

All the data pertaining to the donor and to the lot, in compliance with the legislation, should be kept in single folders and stored in specific files of the musculoskeletal tissue bank for a minimum period of 25 years.

A serum bank with samples of donor plasma should also be made available by the musculoskeletal tissue bank in case of the need for counterproof exams.

As soon as the quality criteria have been evaluated and approved, the tissues are made available for use.

The tissues are distributed to the various specialties (Hip, Knee, Shoulder, Tumors and Dentistry) according to the availability of and requests for grafts.

The transplanter (physician or dentist) places the order for the tissue through a discussion of cases and by sending a specific form. The tissue reservation takes into account the demand for each type of transplant, waiting list and stock. The waiting list for transplantations performed within the Unified Health System - SUS complies with the prevailing legislation and today is organized and managed by the musculoskeletal tissue banks themselves, observing an order by date of inclusion. Urgent cases appointed by the medical team, such as malignant tumors and situations with a risk of severe complications, are communicated to the musculoskeletal tissue bank through an Emergency Form, for immediate response.

In dentistry, transplants have evolved differently and their distribution features some particularities that will be described further on.

12. Global data on tissue transplantation activities in Brazil

Today it can be seen that tissue transplantations in general are on the rise. Events such as officialization in the legislation and the creation of public promotion policies corroborate this evolution. Considering bone transplants alone, 30x growth has been observed in the last 5 years (Brazilian Transplantation Register, 2010), a fact motivated by the start of large-scale distribution of tissues for dental surgery. Although statistics show the number of tissues to be growing, the quantity of donors is still a concern. The vast majority of donors in Brazil are still for the removal of perfused solid organs (heart, kidney, liver, etc. A minority (6% on an average) accept the donation of musculoskeletal tissues. Of these, just 8% on average, become effective donors and the rest are discarded due to the presence of exclusion criteria such as infections, blood transfusion, and inadequate profile (**Graph 1**).

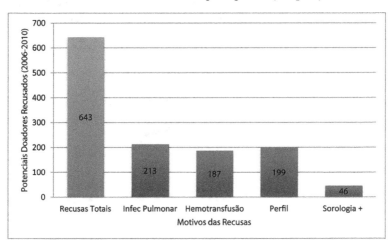

Graph 1. Reasons for refusals of bone donors between 2006-2010.
(Source: File BTME-HC-USP)

13. Homologous tissue transplants in dentistry

Dentistry has also sought biomaterials usable in the replacement of mandibular and maxillary bone loss over the years. The transplantation of bone portions taken from the iliac crest and menton (autologous transplant) has been broadcast in recent years, yet similarly to what happened in the orthopedic area, the disadvantages related to donor morbidity and the good results observed in the use of allografts in orthopedic patients, motivated professionals and patients to adopt this form of treatment.

In Brazil the use of bone transplants in dentistry started in 2005 after a consensus regarding the need for use and administrative mobilizations with the National Transplant System, Federal Council of Dentistry and Musculoskeletal Tissue Banks. This consensus serves as a starting point for the definition of criteria for requests for these tissues at the existing banks, where the professional must be a specialist in the areas of Implant Dentistry, Periodontics or Oral and Maxillofacial.

Rules related to logistics and traceability were incorporated into the activities of existing tissue banks, which then started to implement the bone tissue processing and distribution programs for dental purposes. In this program it is crucial to record the entire process with a focus on health control, including adverse deeds and traceability from the request to the actual transplantation. Specific forms are used for this purpose, including the Request Form, Terminated Transplantation Form, Non-Conformity Term and Adverse Effect Form.

Once standardized, this type of transplant is initiated in the country with widespread adoption by dentists, as observed in the Brazilian transplantation records. (RBT, 2006-2010)

Naturally, the Maxilla and Mandible are today the main bone tissue receptor areas in dentistry that have very distinctive characteristics when submitted to the osteolysis processes, which, in turn, require distinctive techniques during the bone transplant. The size and shape of the bones are influenced by several factors, which range from the genetic conditions of the individual to the environment in which they live. In other words, age, sex, physical characteristics, health, diet, race and place of residence are aspects to be considered (Moore, 1990).

Maxillary and Mandibular development and growth are determined by the appearance of teeth from the first months of life. It is interesting to note that the mandibular and maxillary bone tissue responds to intrinsic and extrinsic factors throughout the lifetime of an individual, and, therefore is very plastic, which counteracts its rigid and inert appearance.

Pathophysiology of Bone Loss
Metabolic Factors: • Age;   • Gender;   • Hormone balance;   • Osteoporosis;   • Nutritional disorders.   *Mechanical Factors:*   • Functional (force applied to edge (pressure, compression, tension, shearing)):

• Frequency; • Direction; • Quantity. *Prosthetic Factors:* • Type of prosthesis base; • Shape and type of teeth.

Table 2. Factors related to the bone loss process (Source: Fonseca & Davis, 1995)

One of the factors most closely related to the indication of bone reconstructions in dentistry are maxillary and mandibular resorptions due to lack of the dental element.

The resorption of the alveolar edge is a chronic, progressive, irreversible and cumulative alteration. This condition, observed in the toothless individual, becomes faster in the first six months after exodontias or dental extractions. Once the function of providing support to the teeth has been lost, the alveolar process tends to undergo resorption due to disuse (Mecall & Rosenfeld, 1991). And this resorption can be exacerbated by local factors (traumatism, infections and pathologies), systemic factors (osteopenia, osteoporosis, osteomalacia, endocrine and nutritional alterations), systemic health problems, prosthetic treatments and others (Fonseca & Davis, 1995; Gassen et al, 2008)

Projections by the Brazilian Institute of Geography and Statistics (IBGE) show that the elderly population in Brazil is set to increase considerably in future years. Life expectancy in 2020 is estimated at 71.2 years (men) and 74.7 years (women) and will represent 13% of the population (IBGE, 2011). Data from the Epidemiological Survey indicate the elderly age bracket as having the highest rates of edentulism and of prosthesis use for prolonged periods. (Ministério da Saúde, 2003)

The previous use not only of total prostheses but also of removable partial prostheses is identified as a predisposing factor of tissue resorption (odds ratio = 2.4), and the flaccid tissue from the edge is related to the severity of resorption (odds ratio = 2.4). (Watzek, 1996, Xie et al, 1997)

In this context, it is noted that the majority of atrophic edentulous cases (total or partial) has increasingly resorted to the adoption of dental implants and for this reason, bone grafts appear as an option of biomaterial used in pre-prosthetic surgery (Galea, 2005).

Frozen homologous bone tissue is biocompatible and can be used successfully in treatments that require maxillary sinus lifting. Its use favors bone neoformation, integration, and absence of inflammatory infiltrate as well as an increase in the percentage of bone volume (Stacchi et al, 2008). In the long term, it is possible to observe the formation of viable and mature bone tissue, providing adequate reconstruction techniques are adopted. (Contar et al, 2009)

Some studies in the dental area have evaluated the efficacy of the use of allografts through an analysis of the biomechanics of implants placed in the grafting zone. This is possible through a resonance frequency analysis (RFA) and removal torque (RT) analysis. The results show that there is no difference in stability between implants installed with autogenous and allogeneic grafts. (Ribeiro, 2009) It is also possible to evaluate the osseointegration process through an evaluation of the neoformed bone volume. Lima (2010) studied homologous bone grafts processed in a tissue bank with different methods (lyophilized, demineralized and radiated (ALD); mineralized frozen (ACM) besides autogenous grafts (AT) and blood clot (CG). In the Guided Bone Regeneration (GBR) technique, samples of the groups of grafts were placed in 32 cylinders fastened to the calvaria of 08 animals. After 13 weeks the cylinder fill rates (bone volume of the ALD group) were similar to the ACM and superior to the autogenous graft). Bone neoformation also occurs during the use of homologous grafts in maxillary sinus lifting surgery, besides affording lower morbidity levels (Viscioni et al, 2010). Hence it should be considered a valid alternative for the replacement of autologous grafts in patients submitted to implant therapy.

14. Prospects in the application of osseointegration investigation methodologies in grafting areas

Some techniques can be used today in the analysis of interface sites between receptor and donor bone, which generate information capable of providing subsidies for a greater understanding of the osseointegration process.

14.1 Histomorphometry

Histomorphometry aims to analyze bone morphology and its components (measurements of volume, area, perimeter etc.). This technique is developed primarily for rock analysis, and is currently employed to analyze cellular behaviors of tissues starting from their structural conformity, expressed in thin slices on a slide. The histological reading and the definition of elements that compose the bone microtexture is the main goal of this procedure. The method can be manual, semiautomatic and automatic. A microscope coupled to a micrometer ruler is used in the first. In the semiautomatic version the microscope is coupled to a computer, which in turn uses software that allows users to record and to quantitatively analyze the images of a slide, seen through the microscope lenses, which are projected onto a digitizing board and drawn manually. The definition of each histological structure is performed by a professional. Finally, the automatic technique allows users to capture the images from the microscope using video cameras and the definition of each structure is determined by the actual computer, which automatically analyzes the coloration of each structure. Although the latter is the fastest technique, it is also the least sensitive. (Jorgetti, 2003; Aaron, Shore, 2003)

This technique permits the analysis of primary or static parameters (extension, number and distance) and the derived parameters, which are divided into structural (analyze bone structure) and kinetic (analyze the dynamics of the bone tissue).

The histomorphometric parameters are measured at 125x zoom. The choice of the analysis area should consider the contact surface of the receptor bone with the allograft.

A microscope equipped with objective, micrometer ruler, cursor, digitizing board and image analysis software is used to quantify the structural static, bone tissue formation and resorption histomorphometric parameters. (**Illustration 22**)

The static histomorphometric parameters are classified as:

STRUCTURAL PARAMETERS

- Total area (**T.Ar** mm^2): total area measured;

- Bone Volume (**BV/TV** %): percentage of bone tissue formed by mineralized or unmineralized trabecular bone;
- Thickness of the woven bone (**Tb.Th** µm): thickness of the bone trabeculae expressed in micra;
- Trabecular number (**Tb.N/mm**): the number of bone trabeculae, by millimeter of tissue, which is also an index that expresses trabecular density;
- Trabecular separation (**Tb.Sp** µm): the distance between the bone trabeculae expressed in micra;
- Number of osteocytes (**N.Ot**): number of osteocytes present in the area of the bone tissue evaluated.

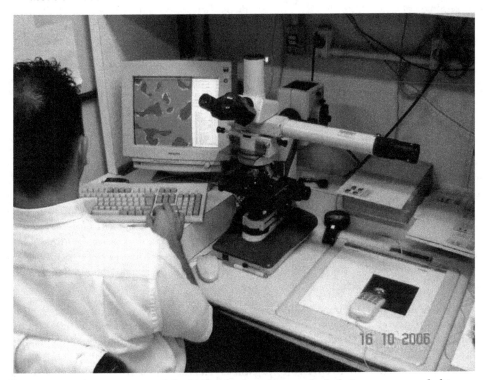

Illustration 14. Equipment used for histomorphometric analysis (microscope coupled to digitizing board and Osteomeasure® software)

FORMATION PARAMETER

- Osteoid volume (**OV/BV** %): percentage of osteoid matrix in relation to the trabecular bone;
- Osteoid surface (**OS/BS** %): percentage of trabecular surface covered by osteoid matrix;
- Osteoblast surface (**Ob.S/BS** %): percentage of the trabecular surface that presents osteoblasts;
- Osteoid thickness (**O.Th** µm): the thickness of the osteoid matrix deposited on the bone trabeculae, expressed in micra;

- Number of osteoblasts (**N.Ob**): absolute number of osteoblasts present in the measured area;
- Number of osteoblasts by tissue area (**N.Ob/T.Ar**): number of osteoblasts by tissue area analyzed, expressed in square millimeters;
- Number of osteoblasts by bone perimeter (**N.Ob/B.Pm**): number of osteoblasts by bone perimeter analyzed, expressed in millimeters.

RESORPTION PARAMETER

- Osteoclast surface (**Oc.S/BS**%): percentage of trabecular surface that presents osteoclasts;
- Number of osteoclasts (**N.Oc**): number of osteoclasts present in the area of the bone tissue evaluated;
- Resorption surface (**ES/BS** %): percentage of surface that presents bone resorption lacunae with or without the presence of osteoclasts.

The histomorphometric parameters adopted in this technique should follow a universal nomenclature agreed upon by the American Society of Bone and Mineral Research - ASBMR (Parfitt et al., 1987).

14.2 Immunohistochemistry

With the advent of osseointegrated dental implants it is extremely important for us to know the cellular processes involved in the osseointegration of allografts, as already mentioned, and this is possible with the immunohistochemistry technique. In vivo mineralization on the implant surface is directly related to the components of the extracellular bone matrix, both collagen and non-collagen components. The bone structure is composed of about 70% of inorganic matter (hydroxyapatite), while the rest is the organic matrix, predominantly consisting of collagen (95%), which is responsible for its flexibility and resistance whereas type I collagen provides support to the mineral structure. The non-collagen components include osteocalcin, osteonectin and osteopontin.

Osteocalcin is related to bone matrix mineralization, and is exclusively expressed by osteoblasts (Ducy et al. 1995). Osteonectin influences the synthesis of extracellular matrix components (Bradshaw et al. 2003) and the calcification of the organic matrix as it is selectively linked to hydroxyapatite and type I collagen fibrils (Termine et al. 1981). Osteopontin is strongly linked to hydroxyapatite crystals and is involved in the anchoring of the osteoclasts in the inorganic bone matrix, controlling crystal nucleation and growth.

Hence the evaluation of these proteins in the osseointegration process can provide us with important information about how the allogeneic bone graft used can corroborate for better interaction of the cellular mechanisms between donor and recipient focusing on the maintenance of the bone tissue after the implant installation.

Moreover, the remodeling of this matrix needs specialized enzymes. Matrix metalloproteinases (MMPs) are an important family of endopeptidases, with 25 known human members, and represent the largest class of enzymes that, collectively, are responsible for the degradation or resorption of extracellular matrix components, and are the only enzymes capable of cleaving fibrillar collagens (Curran et al. 1999), several proteins from the cellular surface and from the pericellular environment and some intracellular proteins.

This technique makes it possible to evaluate bone matrix proteins (type I collagen, osteocalcin, osteopontin and osteonectin) and the MMPs during the osseointegration process of bone grafts in patient to be submitted to future dental implants.

14.3 Peripheral quantitative computed bone tomography (pQCT)

Peripheral quantitative computed tomography - pQCT is a method used to measure the appendicular skeleton that, besides furnishing a broad range of data, also makes it possible to evaluate the cortical and trabecular bones separately.

The obtainment of images is performed by a scanner (ex Stratec XCT Research M instrument scanner (Norland Medical Systems, Fort.) specifically for the analysis of small grafting sites.

The parameters evaluated in this technique are:

- total BMC – total bone mineral content, mg/mm;
- total BMD - total bone mineral density, mg/cm3;
- trabecular BMC - trabecular mineral content, mg/mm;
- trabecular BMD - trabecular mineral density, mg/cm3;
- cortical content, mg/mm;
- cortical density, mg/cm3;
- total area, mm2;
- trabecular area, mm2.
- periosteal circumference, mm;
- endosteal circumference, mm.

15. Discussion

There have been many differences in bone tissue preparation techniques since the first tests in the 30s, yet the desire for the recovery of lost bone tissue among physicians and dentists still remains. The Ministry of Health has recently expressed interest in increasing and promoting this kind of transplant in the country. It is a fact that an increasing number of patients with bone loss for various reasons have approached specialized orthopedics and dentistry services in pursuit of reconstructive solutions. This context includes procedures involving the use of allografts made available by musculoskeletal tissue banks.

In dentistry, the officialization of allograft use arrives very late in 2005. Dental transplanters were curtailed in their use of this kind of treatment for more than a decade. Public policies of oral health were also very late in arriving. The result is perceptible in the statistics that evidence a high rate of edentulism in our population. (Ministerio da Saúde, 2003). In response to this situation, we can also observe high rates of prosthetic rehabilitations in the Brazilian population.

Data from the Epidemiological Oral Health Survey in Brazil (2003) shows that about 67% of the population between 65 and 74 years of age use some kind of prosthesis as a form of edentulism treatment. We know that edentulism and the prolonged use of these prostheses lead to maxillary and mandibular bone resorption over the years (Davis, 1995; Gassen et al,

2008). The high rate of individuals with this necessity, the greater economic access of the population to treatments with osseointegrated implants and to the start of authorization of allograft use in dentistry, corroborated the abrupt growth of bone transplants in dentistry at around **600%** in these last 5 years (**Graph 2**).

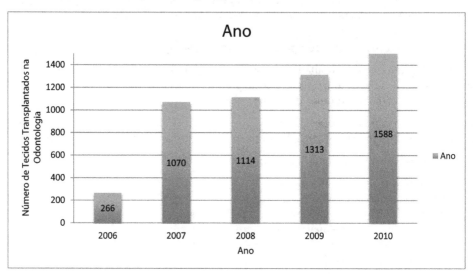

Graph 2. Annual Distribution of homologous tissues made available by a tissue bank that serves as a national reference for Dental Transplants (2006 to 2010).

Such a situation was also observed in our study in relation to the number of transplanting dentists, whose number climbed steeply from **22** to **3585** also in 5 years (**Graph 3**).

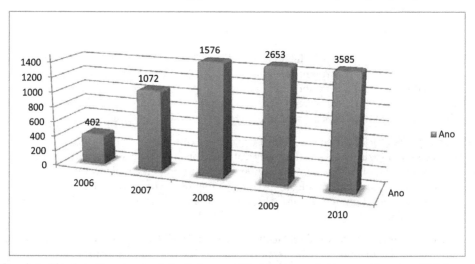

Graph 3. Absolute Frequency of dentists accredited for transplants over the years (2006-2010)

In the next few years we will observe growth of the elderly population in our country (around 13% in 2020), motivated by the increase in life expectancy. It is estimated that in 2020 this expectancy will be 71.2 years for men and 74.7 years for women (IBGE, 2011). Considering that the elderly population is exposed to factors accumulated during their lifetimes that lead to the need for pre-prosthetic reconstructive bone interventions (edentulism, prolonged use of prostheses, osteoporosis, hypothyroidism) we can reflect here on a possible influence over prosthetic rehabilitations of the elderly population in future years, and that in some of them the use of biomaterials (including allograft) may be indicated.

What then has become an obvious trend gives rise to the need to discuss the future of dental transplants. Public health surveillance policies seek, together with trade associations (Regional Federal Council of Dentistry) and tissue banks, the assurance of traceability in procedures of this nature. This is undoubtedly the major challenge for tissue banks. Unlike other kinds of transplants, where the data are compiled and made available by hospitals, in dental transplantations there is the need for total involvement of the transplanting professional in disclosing and informing the adverse effects detected in the treatment. Following the example of the national policy implemented by ANVISA in the area of hemoderivates transfusion, the intention is to create a similar system for the surveillance of tissue transplants in the country, which will include dental transplants. In this model, the adverse effects can be notified nationwide (electronically), enabling the construction of a database that will serve as a source for the definition of corrective and preventive actions of complications in dental surgeries with the use of allografts. Such information is extremely valuable to banks, particularly in the quality control of the tissues produced thereby.

After a more detailed analysis of dental transplants, we detected that approximately 76% are related to the use of cortical tissues, which in our understanding is expected due to the greater availability of this type of tissue at banks. While a spongy segment of a long bone such as the femur and tibia (metaphyseal region) yields 10 to 15 units of spongy blocks, 50 to 60 units of cortical blocks are processed from a diaphysis. It should also be considered that due to the lesser availability of spongy tissues, surgeons have opted in particular for cortical blocks, especially for use in sinus lifting surgery preceding implant installation.

GRAFT MORPHOLOGY	Relative Frequency (%)
Cortical	75.64 %
Spongy	15.86 %
Cortico-Spongy	8.5 %
TOTAL	100.00

Table 3. Distribution of Grafts in Terms of Morphology. BTM (musculoskeletal tissue bank) - HCFMUSP, 2006 to 2010.

The most frequent surgical site was undoubtedly the maxilla (98.2%- Graph 4) and we hereby emphasize the main reasons for this finding. Firstly, edentulism in the Brazilian population has appeared to affect the maxilla more than the mandible which certainly leads to the greater use of upper prostheses (57.9%) versus lower prostheses (34.18%) by the population aged between 65 and 74 years (Ministério da Saúde, 2003). Secondly, the actual anatomy of the maxilla due to the presence of maxillary sinuses that are grafted during the sinus lifting technique. This technique is very frequent in the branch in implant dentistry. Thirdly, the lower bone quality of the maxilla in comparison to the mandible, focusing on the primary stability of implants and propensity for resorption after the prolonged use of prostheses. (Mezzomo et al, 2010) It is also worth emphasizing that the maxilla appears as a more esthetic area that demands more attention from individuals concerned about its rehabilitation.

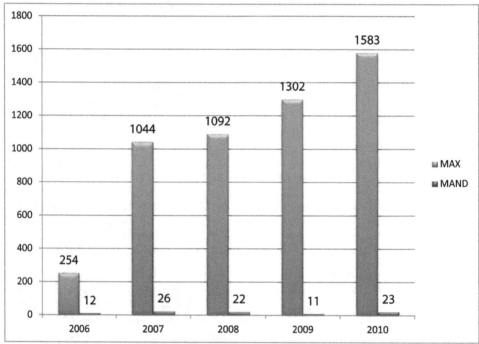

Graph 4. Distribution of Type and Surgical Site of use of the tissues made available by a bank that serves as a national reference from 2006 to 2010.

Although epidemiological data shows the increasing use of homologous tissues in dental surgery, few studies in our field assess their efficacy, hence the need to establish consensuses on their applicability and more importantly, predictability of treatments with the use of allografts. We take predictability to mean knowledge of the responses of bone tissue to the interventions performed in our areas of specialty. Thus in this study we have highlighted a chapter on possible lines of investigation enforceable in our field, which can help us in our search for answers related to the osseointegration process of allografts at sites of dental interest.

The analysis of grafting area by histomorphometry allows us to accurately quantify the structure of the grafted tissue (area, volume) and how much tissue was formed after interaction with the recipient's body (osteoid volume, osteoid surface), to judge the participation of each bone cell (number of osteoblasts and osteocytes) and also to quantify resorptions that have occurred (resorption surface) as a result of this interaction. This data is supplemented by peripheral quantitative computed tomography – pQCT, which unlike histomorphometry, allows us to evaluate the density of the tissue before and after the grafting period. Data of extreme importance, in view of the need for assurance of mechanical stability of dental implants, when implanted in the newly formed bone.

And finally, immunohistochemistry solves some mysteries related to the cellular processes involved in osseointegration. The technique allows us to evaluate, by means of reaction by selected antibodies, the expression of proteins related to the mineralization process (osteocalcin, collagen) and remodeling of bone grafts (osteopontin, osteonectin).

As the basis for all these cellular events, the use of specific antibodies (Tartrate-resistant Acid Phosphatase - TRAP, Osteoprotegerin, Rank-L and Cd34) also allows us to evaluate the inflammation and the vascularization of grafting areas throughout the osseointegration period.

Nowadays in Brazil we are experiencing a time of transition in the indication of these transplants, and studies with this information content are certainly necessary for better knowledge and indication of allografts in dentistry, as well as the criteria and methodologies that can be used to analyze osseointegration.

16. References

[1] Aaron JE, Shore PA. Bone histomorphometry: concepts and commom techniques. In: An YH, Martin KL. Handbook of histology methods for bone and cartilage. New Jersey: Humana Press; 2003. p. 331-51.

[2] Amatuzzi MM, Croci AT, Giovani AMM, Santos LAU, Maragni GG, Shinzato JC. Banco de tecidos:estruturação e normatização. In: Amatuzzi, Joelho:articulação dos membros inferiores. São Paulo:Roca; 2004. p.687-99.

[3] Amatuzzi MM, Croci AT, Giovani AMM, Santos LAU. Banco de tecidos:estruturação e normatização. Rev Bras Ortop. 2000; 35(5):165-72.

[4] American Association of Tissue Banks. Standards for Tissue Banking. 11th ed. McLean : American Association of Tissue Banks; 2007.

[5] Aubin JE, Lian JB, Stein GS. Bone formation: maturation and functional activities of osteoblast lineage cells. In: American Society for Bone and Mineral Research. Primer on the metabolic bone diseases and disorders of mineral metabolism. 6th ed. Washington: ASBMR; 2006. p.20-9.

[6] Bancroft JD, Stevens A. Theory and practice of histological techniques.Fouth Edition. Churchill Livingstone. United Kingdom, 1999.

[7] Barros Filho TEP, Rossi JDMBA, Rodrigues CJ, Lage LA, Reis PR. Poder osteogênico dos enxertos ósseos: estudo experimental comparativo entre enxertos autólogo, homólogo irradiado e homólogo AAA. Rev Bras Ortop. 1989; 24(1-2):36-40.

[8] Bitar AC, Santos LAU, Croci AT, Pereira JARM, França Bisneto EN, Giovani AMM, Oliveira CRGCM. Histological Study of Fresh versus Frozen Semitendinous Muscle Tendon Allografts. CLINICS 2010;65(3):297-303

[9] Boldt JG, Dilawari P, Agarwal S, Drabu KJ. Revision total hip arthroplasty using impaction bone grafting with cemented nonpolished stems and charnley cups. J Arthroplasty. 2001; 16(8):943-52.

[10] Bradshaw, A.D., Graves, D.C., Motamed, K., & Sage, E.H. 2003. SPARC-null mice exhibit increased adiposity without significant differences in overall body weight . Proc.Natl.Acad.Sci.U.S.A, 100, (10) 6045-6050 available from: PM:12721366

[11] Brasil, Leis etc. Decreto n.2268 de 30 de junho de 1997. Dispõe sobre a remoção de órgãos, tecidos e partes do corpo humano para fins de transplantes e tratamento. Diário Oficial da União, Brasília (DF). 1997 30 jun; seção 1:1.

[12] Brasil, Leis etc. Lei n.9434 de 5 de fevereiro de 1997. Dispõe sobre a remoção de órgãos, tecidos e partes do corpo humano para fins de transplantes e tratamento. *Diário Oficial da União*, Brasília (DF). 1997 5 fev; seção 1:25.

[13] Brasil, Leis etc. Portaria n.1686 de 20 de setembro de 2002. Dispõe sobre a regulamentação para funcionamento de banco de tecidos músculo esqueléticos. *Diário Oficial da União*, Brasília (DF). 2002 24 jul; seção 1:1.

[14] Brasil, Leis etc. PORTARIA Nº 2.600, DE 21 DE OUTUBRO DE 2009. Aprova o Regulamento Técnico do Sistema Nacional de Transplantes. *Diário Oficial da União*, Brasília (DF). 2009 21out.

[15] Brasil, Leis etc. Resolução n. 220 de 27 de dezembro de 2006. Dispõe sobre o Regulamento Técnico para o Funcionamento de Bancos de Tecidos Musculoesqueléticos e de Bancos de Pele de origem humana. *Diário Oficial da União*, Brasília (DF). 2006 29 dez.

[16] Cabrita, H.B., et al., Prospective study of the treatment of infected hip arthroplasties with or without the use of an antibiotic-loaded cement spacer. Clinics (Sao Paulo), 2007. 62(2): p. 99-108.

[17] Carrel A. The preservation of tissues and its applications in surgery. 1912. *Clin Orthop Relat Res.* 1992; 278:2-8

[18] Contar CM, Sarot JR, da Costa MB, Bordini J, de Lima AA, Alanis LR, Trevilatto PC, Machado MÂ. Fresh-frozen bone allografts in maxillary ridge augmentation: histologic analysis. J Oral Implantol. 2011 Apr;37(2):223-31.

[19] Croci AT, Camargo OP, Oliveira CRGC, Baptista AD, Sorrilha A. Estudo histológico dos enxertos ósseos homólogos humanos. *Acta Ortop Bras.* 2003b; 11(4):220-4.

[20] Cunningham N, Reddi AH. Biologic principles of bone induction: application to bone grafts. In: Habal MB, Reddi AH. Bone grafts and bone substitutes.Philadelphia: W.B . Saunders C; 1992. p.93-8.

[21] Curran, S. & Murray, G.I. 1999. Matrix metalloproteinases in tumour invasion and metastasis. *J.Pathol.*, 189, (3) 300-308 available from: PM:10547590

[22] Drumond SN. Transplantes ósseos. In: Pereira WA. *Manual de transplantes de órgãos e tecidos.* 2a ed. Rio de Janeiro: Medsi; 2000. p.359-80.

[23] Ducy, P. & Karsenty, G. 1995. Two distinct osteoblast-specific cis-acting elements control expression of a mouse osteocalcin gene *Mol.Cell Biol.*, 15, (4) 1858-1869 available from: PM:7891679

[24] Enneking WF, Burchardt H, Puhl JL, Piotrowski G. Physical and biological aspects of repair in dog cortical-bone transplants. *J Bone Joint Surg Am.* 1975; 57:237-52.

[25] European Association of Tissue Banks. Common Standards for Tissues and Cells Banking: Berlin: European Association of Tissue Banks; 2004.

[26] Fischer LP, Fischer-Athiel C, Fischer BS. One hundred years of bone surgery in the Lyons Teaching Hospitals (1897-1997). *Ann Chir.* 1998; 52(3):264-78.

[27] Fonseca RJ, Davis WH .Reconstructive Preprosthetic Oral and Maxillofacial Surgery, ed 2. Philadelphia, WB Saunders, 1995, pp 498-510

[28] Galea G, Kearney. Clinical effectiveness of processed and unprocessed bone. *Transfus Med.* 2005; 15(3):165-74.

[29] Gassen, Humberto Thomazi; Filho, Rolf Muner; Siqueira, Bianca Munari de; Oliveira, Samia Bohm; Junior, Aurelício Novaes Silva. Reconstrução óssea de maxila atrófica utilizando enxerto de ramo mandibular. Stomatos, v.14, n.26, jan./jun. 2008.

[30] Gerstenfeld LC, Einhorn TA. Fracture healing: the biology of bone repair and regeneration. In: American Society for Bone and Mineral Research. Primer on the

metabolic bone diseases and disorders of mineral metabolism. 6th ed. Washington: ASBMR; 2006. p.42-48.

[31] Giannoudis PV, Dinapoulos H, Tsiridis E. Bone substitutes: an update. Injury. 2005; 36S:S20-27.

[32] Giovani AMM, Croci AT, Oliveira CRGCM, Filippi RZ, Santos LAU, Maragni GG. Comparative study of cryopreservedbone tissue and tissue preserved in a 98% glycerol solution. Clinics. 2006;61(6):565-70.

[33]Giovani AMM. *Estudo comparativo entre o tecido ósseo criopreservado e o conservado em glicerol a 98% [dissertação]*. São Paulo: Faculdade de Medicina, Universidade de São Paulo; 2005.

[34] Groves EWH. Methods and results of transplantation of bone in the repair of defects caused by injury or disease. *Br J Surg.* 1917; 5(18):185-242.

[35] Heyligers IC, Klein-Nulend J. Detection so living cells in non-processed but deep-frozen bone allografts. *Cell Tissue Bank.* 2005; 6 :25-31.

[36] IBGE: Instituto Brasilerio de Geografia e Estatística: http://www.ibge.gov.br/home/ . Acessado em 15 de julho de 2011.

[37] Janssen ME, Lam C, Beckham R. Outcomes of allogenic cages in anterior and posterior lumbar interbody fusion. *Eur Spine J.* 2001; 10(Suppl 2):158-68.

[38] Jorgetti V. Biópsia óssea e análise histomorfométrica. Manual Fleury de diagnóstico de doenças ósteo-metabólicas. 2003. Disponível em: http://www.fleury.com.br/htmls/cdrom/capitulo5.htm.

[39] Junqueira LC, Carneiro J. Histologia básica. 9a ed. Rio de Janeiro :Guanabara-Koogan; 1999. Cap.8, p.111-28: Tecido ósseo.

[40] Knighton DR, Hunt TK, Thakral KK, Goodson WH 3rd. Role of platelets and fibrin in the healing sequence: an in vivo study of angiogenesis and collagen synthesis. *Ann Surg.* 1982; 196 (4): 379-88.

[41] Lavernia CJ, Malinin TI, Temple HT, Moreyra CE. Bone and tissue allograft use by orthopaedic surgeons. *J Arthroplasty.* 2004; 19(4):430-35.

[42] Lindhe J, Karring T, Lang NP.Tratamento periodontal regenerativo. Tratado de periodontia clínica e implantologia oral. 3a ed. Rio de Janeiro: Guanabara Koogan;1997. p.428-62.

[43] Malafaya, PB, G. A. Silva, E. T. Baran, R. L. Reis, Curr.Opin. Solid State Mater. Sci. 2002, 6, 283.

[44] Mecall RA, Rosenfeld AL. The influence of residual ridge resorption patterns on implant fixture placement and tooth position. Part I. Int J Periodontics Restorative Dent. 1991; 11(1): 8-23

[45] Mezzomo RJ, Garbin CA, Schuh C, Rigo L. Critical analysis of studies of immediately loaded implants supporting fixed prostheses in the edentulous maxilla. Implantnews 7 (4), 2010.

[46] Ministério da Saúde : Levantamento das condições de saúde bucal da população brasileira; 2003. Disponível em : http://dab.saude.gov.br/CNSB/vigilancia.php

[47] Moore KL. Anatomia Orientada para a Clínica. 2ª ed. Rio de Janeiro - RJ: Editora Guanabara. 1990.

[48] Nather A. Organisation, operational aspects and clinical experience of National University of Singapore Bone B. *Ann Acad Med Singapore.* 1991; 20(4):453-7.

[49] Parfitt AM, Drezner MK, Glorieux FH, Kanis JA, Malluche H, Meunier PJ, et al. Bone histomorphometry: standardization of nomenclature, symbols, and units. Report of the ASBMR Histomorphometry Nomenclature Committee. *J Bone Miner Res.* 1987; 2(6):595-610

[50] Perren SM, Claes L. Biologia e biomecânica no manejo de fraturas. In: Rüed TP, Murphy WM. *Princípios AO do tratamento de fraturas*. Tradução de Jacques Vissoky.Porto Alegre: Artmed; 2002. p.7-30.

[51] Phillips GO, Strong DM, Versen RV, Nather A. Advances in Tissue Banking. Vol. 4. Worl Scientific . New Jersey, 2000.

[52] RBT- Registro Brasileiro de Transplantes [on line] São Paulo: Associação Brasileira de Transplantes de Órgãos; 2006-2010. Disponível em: http://www.abto.org.br/profissionais/profissionais.asp.

[53] Santos LAU. Efeito da utilização de plasma rico em plaquetas na osteointegração dos enxertos ósseos homólogos criopreservados : estudo histomorfométrico em coelhos [dissertação]. São Paulo: Faculdade de Medicina, Universidade de São Paulo; 2007. Disponível em: http://www.teses.usp.br/teses/disponiveis/5/5140/tde-16082007-160750/

[54] Santos, LAU et al .Banco de Tecidos Musculoesquelético: Da captação ao uso clínico. In: Mancussi e Faro, AC. Emergencias ortopédicas. São Paulo. Editora Manole.304p.2011

[55] Schreurs BW, Busch VJ, Welten ML, Verdonschot N, Slooff TJ, Gardeniers JW. Acetabular reconstruction with impaction bone-grafting and a cemented cup in patients younger than fifty years old. *J Bone Joint Surg Am*. 2004; 86(11):2385-92.

[56] Sharrard WJW, Collins DH. The fate of human decalcified Bone grafts. *Proc R Soc Med*. 1961; 54:1101-2.

[57] Smith SE, DeLee JC, Ramamurthy S. Ilioinguinal neuralgia following iliac bone-grafting. Report of two cases and review of the literature. *J Bone Joint Surg Am*.1984;66(8):1306-8.

[58] Stacchi C, Orsini G, Di Iorio D, Breschi L, Di Lenarda R. Clinical, histologic, and histomorphometric analyses of regenerated bone in maxillary sinus augmentation using fresh frozen human bone allografts. J Periodontol. 2008 Sep;79(9):1789-96.

[59] Thies RS, Bauduy M, Ashton BA, Kurtzberg L, Wozney JM, Rosen V. Recombinant human bone morphogenetic protein-2 induces osteoblastic differentiation in w-20-17 stromal cells. *Endocrinology*. 1992; 130(3):1318-24.

[60] Tomford WW, Mankin HJ. Bone Banking: Update on methods and materials. *Orthop Clin North Am*. 1999; 30:565-70.

[61] Tomford WW. Bone allografts: past, present and future. *Cell Tissue Bank*. 2000; 1:105-9.

[62] Tuominen T. *Native bovine bone morphogenetic protein in the healing of segmental long bone defects* [dissertação]. Oulu: Division of Orthopaedic and Trauma Surgery, University of Oulu; 2001.

[63] Urist MR, DeLange RJ, Finerman GA. Bone cell differentiation and growth factors. *Science*.1983; 220(4598):680-6.

[64] Urist MR. Bone formation by auto induction. *Science*.1965; 50:893-99.

[65] Viscioni A, Dalla Rosa J, Paolin A, Franco M. Fresh-frozen bone: case series of a new grafting material for sinus lift and immediate implants. J Ir Dent Assoc. 2010 Aug-Sep;56(4):186-91

[66]Watzek G. Endosseous implants:scientific and clinical aspects. Chicago; Quintessence; 1996.

[67] Weyts FA, Bos PK, Dinjens WN, van Doorn WJ, van Biezen FC, Weinans H, Verhaar JA. Living cells in 1 of 2 frozen femoral heads. *Acta Orthop Scand*. 2003; 74(6): 661-4.

[68] Xie Q, Narthi T. Oral status and prosthetic factors related to residual ridge resorption in elderly subjects. Acta Odonto Scand 1997; 55:306-13.

2

Cryopreservation of Rat Bone Marrow Derived Mesenchymal Stem Cells by Two Conventional and Open-Pulled Straw Vitrification Methods

Mohammad Hadi Bahadori[*]

Cellular and Molecular Biology Research Center, Faculty of Medical Science,
Guilan University of Medical Sciences, Rasht,
Iran

1. Introduction

Bone marrow (BM) is a complex tissue containing populations of progenitor and stem cells (1). One type, hematopoietic stem cells (HSCs), can renew circulating blood elements such as red blood cells, monocytes, platelets, granulocytes and lymphocytes. The other is mesenchymal stem cells (MSCs), which possess two important properties of long-term self renewal and differentiate into osteoblasts, chondroblasts, adipocytes and hematopoiesis supporting stroma (2, 3). Their mesenchymal differentiation potential is retained even after repeated subcultivation *in vitro* (4, 5). Besides originating the forming mesenchymal tissue, many studies have demonstrated that MSCs could differentiate into various non-mesenchymal tissue lineages under appropriate experimental conditions *in vitro* and *in vivo*, such as hepatocytes (6, 7), cardiomyocytes (8, 9), lung alveolar epithelium (10), olfactory epithelium (11), inner hair cells (12), neurons and neuroglia (1, 4, 13). MSCs are spindle shaped fibroblast-like cells that are easily isolated, cultured and expanded *in vitro* due to their adherent characteristics, and not associated with any ethical debate (14). Thus, MSCs may be used in the treatment of a diverse variety of clinical conditions (15) such as engraftment of various organs (16). The long-term cultivation of MSCs may fail for many reasons: genotypic drift, senescence, transformation, phenotypic instability, and contamination or incubator failure. The inability to cultivate MSCs will result in the lack of MSCs for both experimental and clinical use (17). Therefore, it is necessary to cryopreserve MSCs as cell seeds. Although increasing telomerase expression of cells may overcome cell senescence (18), cryopreservation of hMSCs may be more practical in order to save time and culture materials (16, 19). Resuscitated MSCs can be subcultivated for many passages without a noticeable loss of viability and capability of osteogenic differentiation (20-22).

Formulating a cryopreservation protocol for hMSCs is required because these cells cannot survive for long periods under *in vitro* culture conditions. Slow rate cooling methods using dimethylsulfoxide (DMSO) as a cryoprotectant have been used for a wide variety of MSC lines established from bone marrow (23, 24), umbilical cord blood (23-25), hematopoietic progenitor cells (26) and mouse ES cell lines (27). In most protocols, cells are suspended in

---

[*] Corresponding Author

freezing medium containing DMSO at 5-20%, transferred into glass or plastic cryovials and then frozen by cooling at 1.0 to 2.0 °C/min (28). Slow freezing reduces ice crystal formation and eliminates toxic and osmotic damage to cells through exposure to low concentrations of cryoprotectants while slowly decreasing temperatures (29). However, it is difficult to completely eliminate injury by intracellular ice formation. Damage by ice crystal formation in the cytoplasm during the freezing process is one of the possible causes of cell death; such conventional methods, are not applicable to hMSC cells because many of these cells die immediately after thawing (28). Alternately, vitrification, a rapid cooling method using a high concentration of cryoprotectant, could also be used. Vitrification can completely eliminate damage caused by ice crystal formation in the cytoplasm of cells during freezing (29, 30). It is also advantageous because the procedure takes a relatively short time and a programmable temperature decreasing container is not required (31).

Vitrification has been used for the cryopreservation of oocytes, fertilized eggs and embryos of several mammalian species including humans in order to prevent ice crystal formation (32). There have been some reports demonstrating that embryonic stem (ES) cells could be successfully cryopreserved by vitrification in recent years (27, 33,34). Moon et al. tested vitrification of the human amnion-derived mesenchymal stem cells (HAMs) by using a two-step exposure to equilibration and vitrification solutions (21). They used an EG-based cryoprotectant and their findings were in line with previous reports that showed the superiority of EG. EG has been proven to be less toxic on fibroblast and other somatic cells in comparison with permeating agents such as DMSO and propylene glycol (PROH) that have been used on murine and human embryos (35). However, as a long-term preservation method for HAMs, a well-defined protocol of cryopreservation needs to be established for a human bone marrow derived mesenchymal stem cell bank. In the present study, to confirm the proliferative capability and pluridifferentiation of cryopreserved adult hMSCs; we chose ethylene ficoll sucrose (EFS) 40 that contained 40% v/v EG for the vitrification solution. hMSCs that were cryopreserved for two months were resuscitated and cultivated for 15 passages. An analysis of their expansion, morphological and pluridifferentiation characteristics was undertaken. Finally, under induction conditions, adipogenic and osteogenic potentials have been discussed.

2. Materials and methods

2.1 Preparation and culture of MSCs

For isolation of rat MSCs; female Sprague-Dawley rats (weighing 200-250 g) with the approval from the Institute for Animal Care were obtained from the Animal Center, Faculty of Medicine, Guilan University of Medical Sciences. Rats were killed by intraperitoneal administration of a lethal dose of sodium pentobarbital. The femurs and tibias were carefully dissected away from attached soft tissue as previously reported with modification (1). The ends of the bones were cut, and the bone marrow was aseptically extruded with 5 ml PBS solution by using a syringe with a 21G needle and flushing the shaft ten times. The marrow tissue was dissociated by pipetting. The cell suspension was then centrifuged at 500 × g for 5 minutes and the supernatant was discarded. Bone marrow mesenchymal stem cells (BMSCs) were then mechanically dispersed into a single-cell suspension so that the density of BMSCs reached 10^6 cells/ml. At this point, marrow cells were plated in a 25 cm^2 plastic flask in Dulbecco's modified eagle medium (DMEM) containing 20% fetal bovine serum

(FBS), 100 U/ml penicillin, and 100 mg/ml streptomycin. All cells were incubated at 37 °C, in an atmosphere of 5 % humidified CO_2. After 48 hours incubation, the nonadherent cell populations were removed and the medium was added and replaced every three or four days for about two weeks. When the cells grew to 80% confluency they were harvested with 0.25% trypsin and 1 mM EDTA (Gibco, UK) for 5 minutes at 37°C, replated and diluted 1:3 on a 25 cm^2 plastic flask, again cultured to the next confluency and harvested. Prior to their use in inducing differentiation and vitrification MSCs that were passaged approximatly 15 times were morphologically evaluated.

2.2 Cryopreservation of MSCs

MSCs at passage 4 of pre-cryopreservation were harvested and centrifuged at 400 × g for 15 minutes as mentioned above. Approximately 1 ×10^6 cells/ml of randomly selected batches were cryopreserved by using the vitrification method or OPS vitrification.

2.3 Vitrification and thawing procedure

MSCs were cryopreserved by using a two-step exposure to the equilibration and vitrification solutions (34). The equilibration solution was 20% ethylene glycol (EG; Sigma) and the vitrification solution was composed of 40% EG, 18% Ficoll - 70 (Sigma) and 0.3 M sucrose (Sigma). All solutions were based on PBS (Sigma) containing 20% FBS. A pellet of ~1 × 10^6 MSCs (~10 μl) was first suspended in 50 μl equilibration solution for 5 minutes and then mixed with 500 μl vitrification solution for 40 seconds. Suspended MSCs were immediately transferred to 1.2 ml cryovials (Nunc) and plunged directly into liquid nitrogen. The OPS vitrification method was carried out according to Reubinoff et al. (33). For OPC, a pellet of ~1 × 10^6 MSCs (~10 μl) was first suspended in 50 μl equilibration solution for 5 minutes and then mixed with 500 μl vitrification solution for 40 seconds. Suspended MSCs were at once transferred to 0.25 ml plastic straws (IMV, L'Aigle, France). Immediately afterwards, the straws were immersed in liquid nitrogen for two months. Following storage, the cells were thawed by rapidly immersing the vials and straws in a water bath at 37 °C. After warming for about 7 seconds, (at approx. 1800 °C/min) the contents of the vials and straws were suspended serially in 0.5, 0.25 and 0 M sucrose in PBS containing 20% FBS. After thawing, the survival rate was evaluated by the trypan blue staining method. After removing some of the cell pellet and adding 0.4% trypan blue (Sigma), the cells were plated onto a slide and unstained cells were counted as live cells (26). The remaining cells were centrifuged at 200 × g for 10 minutes and washed three times with DMEM medium supplemented with 20% FBS, 100 U/ml penicillin and 100 mg/ml streptomycin. Cells were immediately plated at a density of 1 × 10^6 cells/ml in a 25 cm^2 culture flask and subcultured over seven days in the above described condition.

2.4 Evaluation of the differentiation potential of cryopreserved MSCs

2.4.1 Adipogenic induction

Pre and post-cryopreserved MSCs were seeded on coverslips in a six-well plate and cultured in DMEM with 10% FBS. Cells with nearly 80% confluency were exposed to DMEM supplemented with 5μg/ml insulin, 1 μM dexamethasone, 100 nM indomethacine, 0.5 mM methylisobutylxanthine (Sigma), and 10% FBS for 48 hours. Cells were then incubated in the same medium without dexamethasone. For control, cells were cultured in regular medium as above. The medium was changed every third or fourth day. One week after induction, adipogenic differentiation was evaluated by the cellular accumulation of neutral lipid vacuoles that were stained with oil-red O (Sigma) and observed under an inverted microscope (17). Briefly, after fixation in 5% metanol, induced MSCs were stained in filtered oil red O for 2-3 hours and then rinsed with 60% isopropyl alcohol.

2.4.2 Osteogenic induction

To identify osteogenic differentiation, thawed and non-cryopreserved MSCs were cultured in 100 nM dexamethasone, 10 mM ß-glycerol phosphate and 50 µM ascorbic acid-2-phosphate in 400 µl DMEM-LG supplemented with 10% FBS on coverslips in a six-well plate for subsequent staining. During the culture period, the medium was changed once per week. After 14 days, osteogenic differentiation was evaluated by staining the coverslips with fresh 0.5% alizarin red solution (1).

2.4.3 Colony-forming unit assays

For these assays, both thawed and non-cryopreserved cells were plated at 1×10^6 cells per ml and cultured for 14 days in 25 cm^2 tissue culture flasks. After 14 days, the cultures were stained with giemsa for 5 minutes. The formations of colonies were considered acceptable until passage 15 (P15) and those less than 2 mm in diameter or faintly stained were excluded.

2.4.4 Statistical analysis

Statistical analysis for comparison of the postthaw survival rate was performed using the χ^2 test. Statistically significant values were defined as $p<0.05$. All experiments were conducted in triplicate.

3. Results

The growth and morphology of MSCs appeared rather heterogeneous in primary culture as seen in Fig 1A. Under a phase contrast microscope, the cells appeared fibroblast-like, elongated and spindle shaped with a single nucleus (Fig 1B). These cells showed the ability to form colonies with the occasional cell sphere formation giving an impression of embryoid bodies (Fig 1C). However, they progressively showed homogenous fibroblast-like features following subsequent subculture (Fig 1D).

3.1 Morphology and growth of vitrified-thawed MSCs

The duration of storage in frozen state for MSCs was two months. Post-cryopreserved MSCs from both the vitrification method and OPS vitrification had similar cellular morphology and colony-formation. Resuscitated MSCs first grew as isolated colonies after initial plating. Subsequently these adherent cells grew as typically fibroblastic or spindle shaped.

As the cells approached confluence, they assumed a more spindle-shaped, fibroblastic morphology (Fig 2A, B). The thawed and non-cryopreserved MSCs were subcultured until P15. Until the P9, fibroblast-like morphology was consistently observed in both the thawed and non-cryopreserved MSCs. At the P10, cells in both cultures became large and flat, suggesting senescence.

Cryopreservation of Rat Bone Marrow Derived Mesenchymal Stem Cells by Two Conventional and Open-Pulled Straw Vitrification Methods

39

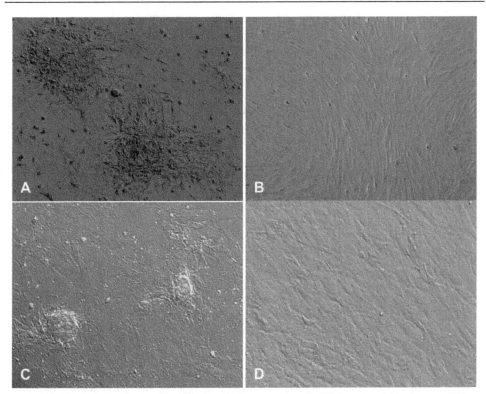

Fig. 1. Morphology and growth of MSCs. (A) Primary (x 40), (B) Passage 4 (14 days, x 100), (C) Passage 4 (35 days, x 100) and (D) Passage 4 (40 days, x 100). In primary culture, cell growth was scattered with some colony formation. Followin subsequent subculture, the cells changed into spindle-like fibroblasts.

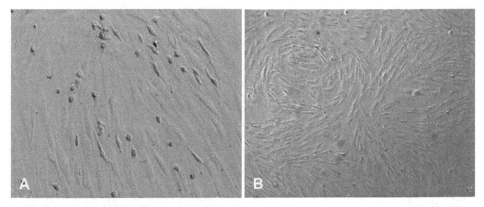

Fig. 2. Morphology and growth of vitrified-thawed MSCs. Phase contrast images of MSCs two months after thawing from: (A) vitrification method (x 100) and (B) OPS vitrification (x 40). MSCs had a similar morphology to fibroblasts and were indistinguishable from non-cryopreserved MSCs.

4. Viability of vitrified-thawed MSCs

Live/dead viability of MSCs was determined by the trypan blue staining test. The number of MSCs was counted and compared to that of the control. After thawing, the viability rates were 81.33% ± 6.83 for the vitrification method and 80.83% ± 6.4 for the OPS vitrification, while values with the pre-vitrification control group were 88.16% ± 6.3, respectively (Table 1). There were no differences in viability between them.

No.	Before vitrification (%)	Vitrification method (%)	OPS method (%)
1	85	81	82
2	93	92	90
3	78	77	70
4	96	81	79
5	89	72	81
6	88	85	83
Mean ± SD	88.16(SD ± 6.30)	81.33 (SD ± 6.83)	80.83(SD ± 6.40)

OPS: open-pulled straw, student t-test (p<0.05).

Table 1. Percentage of cell viability of non-cryopreserved and two different cryopreserved vitrification methods by the trypan blue staining test.

5. Colony forming unit assay

For these assays, cells of both thawed and noncryopreserved MSCs or cultures were plated at 1× 10^6 cells/ml and MSCs in culture showed a colony formation consisting of 40-80 cells in 25 cm^2 flasks for 14 days (Fig 3). After P9, the cells showed no colony forming ability which illustrated that colony forming ability decreases with increasing passages.

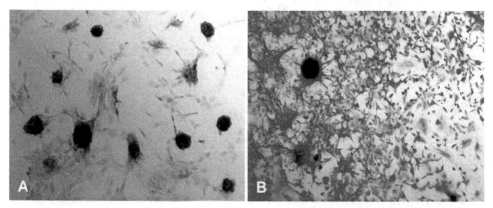

Fig 3. Phase contrast images of typical MSCs colony morphology before (A) and after (B) cryopreservation. Cell sphere formation in the MSCs culture produced colony formation units that stained with giemsa (A and B: x 40).

Cryopreservation of Rat Bone Marrow Derived Mesenchymal Stem Cells by Two Conventional
and Open-Pulled Straw Vitrification Methods

41

6. Differentiation of post-cryopreserved MSCs

After culturing for adipogenic differentiation, the accumulations of numerous neutral lipid vacuoles were detectable in the cytoplasm of vitrified-thawed cells (Fig 4A). Following three weeks of induction, oil Red O staining showed the lipid droplets with orange red color, which demonstrated the committed differentiation of MSCs into adipocytes (Fig 4B). The control cells showed no detectable lipid vacuoles. Under culture with osteogenic induction medium, resuscitated MSCs detached and floated in the medium. After three weeks, mineral accumulations were observed by alizarin red staining (Fig 5A, B). Similar results were observed in the group of non- cryopreserved MSCs when osteogenically induced under the same condition (Fig 5C, D).

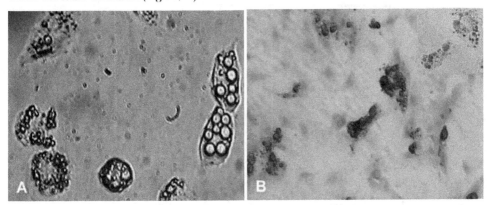

Fig 4. Evaluation of adipogenesis potential of MSCs under a phase-contrast microscope. Both the non cryopreserved MSCs and vitrified-thawed MSCs after treatment by adipogenic medium showed numerous neutral lipid vacuoles which accumulated in the cytoplasm. (A) Confirmed by oil red O staining (oil red O + hematoxiline, which one is the top right one: oil red without hematoxiline, x 100), (B).

7. Discussion

Cryopreservation is an important method to maintain cells for biological research and medical applications such as tissue engineering, gene therapy, cell transplantation, pharmacological testing and future therapeutic indications (17, 28). A study on the long-term storage of BM-derived MSCs is of critical importance (1). The objective of the current investigation was to test the possibility that vitrification could be a useful method for the cryopreservation of MSCs. Thus, in the present study, we isolated MSCs from bone marrow of adult female rats. In culture; MSCs are characterized by their capacity to adhere to a plastic culture surface and form a fibroblast-like shape (Fig 1). Our data corroborated previous findings from other groups which showed homogenous fusiform features with oval vesicular nuclei (36) and the colony forming ability of MSCs, which decreased with increasing passages (37). The achievement of pure fibroblastic clones from murine bone marrow was first reported by Wang and Wolf (38).

Furthermore, Eslaminejad et al. obtained an average of 15-17 clones, each one consisting of several fibroblastic cells per 24-well plate (39). This approach yielded both the number and

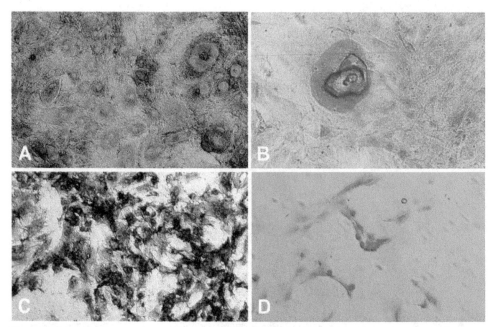

Fig 5. Differentiation potency of nonvitrified and vitrified-thawed MSCs observed under a phase-contrast microscope. (A) Nonvitrified MSCs after osteogenic induction. (B) Alizarin red staining, x 100. (C) Vitrified MSCs after osteogenic induction. (D): Alizarin red staining, x 100. After induction of differentiation for 2-3 weeks in respective induction media, the mineralized extracellular matrix of pre and post-vitrification MSCs stained positively with alizarin red (arrows).

cellular densities of colonies that were dependent upon the number of the cells plated per culture dish and was the method employed in this investigation. We also successfully vitrified MSCs and confirmed the morphology, viability rate and differentiation capacity of post-vitrification MSCs as compared with non-cryopreserved controls. Generally, most comprehensive studies on the cryopreservation of MSCs were carried out by using slow-rate cooling methods (17, 22-24). Slow-rate cooling methods using dimethylsulfoxide (DMSO) as a cryoprotectant are effective for a wide variety of cell lines, including ES cell lines (27, 28). This method by decreasing temperature slowly with a low concentration of cryoprotectant is used to balance damage caused by various factors including ice crystal formation, fracture, toxic and osmotic damage (29).

It is difficult to completely eliminate injuries from intracellular ice formation, which is the main source of fracture and damage to the cytoplasm (29). It is also a time-consuming procedure and requires an expensive programmable freezer (31). On the other hand, there are some reports that stem cells are highly sensitive to cryoinjury and the vitrification method is a better choice for HES cell cryopreservation than conventional slow freezing and rapid thawing (28, 33). This method has been previously applied for cryopreservation of oocytes, fertilized eggs and embryos of several mammalian species, including humans (32).

Cryopreservation of Rat Bone Marrow Derived Mesenchymal Stem Cells by Two Conventional
and Open-Pulled Straw Vitrification Methods

43

In vitrification method, the concentrations of cryoprotectants seem to be dangerously high at the final phases, it happens at low temperatures, where the real toxic effect is minimal. Moreover, the high cooling and warming rates applied at vitrification provide an unique benefit compared to the traditional freezing (40).

Umbilical cord blood-derived mesenchymal stem cells (UCB-derived MSCs) were vitrified by vitrification and by programmed freezing without dimethyl sulfoxid (DMSO) by Wang et al. (2011). Their results showed that the viability of thawed UCB-derived MSCs was enhanced from 71.2% to 95.4% in the presence polyvinyl alcohol (PVA) for vitrification, but only < 10% to 45% of viability was found for programmed freezing (41). While, Kim et al. reported that Post-thaw colony-formation of embryonic stem cells (ESCs) was detected only after a slow freezing using DMSO by stepwise placement of a freezing container into a -80°C deep freezer and subsequently into -196°C liquid nitrogen, while no proliferation was detected after vitrification (42). Also, hMSCs from pre- and post-cryopreservation by slow freezing had similar colony-formation and cellular morphology similar to our resuls (17). Low survival of human ESCs has been also reported when they are frozen slowly with DMSO (43). There have been several reports to demonstrate the superiority of vitrification to other freezing programs for human ESCs, because it is able to avoid cell injury resulted from ice crystal formation (42, 44).Carvalho et al. (2008) reported that viability of frozen BM-MSCs by slow freezing which was to added 5% DMSO was 94.76% and 90.58% viability before versus after cryopreservation (45). Also, the high survival rate (81.8%) is obtained after cryopreservation of human ESCs by programmed freezing (46). Our result showed 81.33% and 80.83% cell viability of two different cryopreserved vitrification methods versus 88.16% viability of non-cryopreserved cells. Nevertheless, we are not able to deny the feasibility of vitrification program for effective cryopreservation of SCs (42).

Also Fujioka et al. vitrified ESCs using EFS40; ethylene glycol, 18% ficoll 70,000 MW and 0.3 M sucrose (similar to this work) and slow-freezed in freezing medium containing 10% DMSO. Their results indicated that the vitrification methods yielded higher cell recovery and survival rates than did the slow-rate freezing methods (28).

DMSO has been well known as a toxic agent for stem and progenitor cells, and particularly for human embryonic stem cells. It is also known as a powerful differentiation agent that may interact with the chromatin structure (47).

Therefore, the selection of suitable cryoprotectants is essential. Cryopreservation procedure ought to become different according to cell type and cellular characteristics. So, all components consisting of freezing and thawing procedures should specifically be determined in each case of cryopreservation (42).

This study was DMSO-free vitrification of MSCs using cryovial and straw. Straws, preferably thin straws, even in sealed form, can be cooled safely with an increased cooling rate (40). There was any report about vitrification of MSCs without DMSO.

So that in the present experiment, we have observed the viability and proliferation capability as well as differentiation potential of cryopreserved MSCs *in vitro*. Inverted microscope findings showed that, after culturing for seven days, numerous MSCs adhered well to the surface culture dish (Fig 1). In a study by Moon et al. on human amnion-derived mesenchymal stem cells (IIAMs), they observed that slow freezing resulted in a lower survival rate compared with vitrification, indicating a high efficiency of the vitrification procedure (21).

Moreover, Heng (48) demonstrated that $39.8 \pm 0.9\%$ of the hMSCs could be recovered after cryopreservation using a conventional slow freezing method which was lower than that our result (Table 1). Here, resuscitated MSCs kept a high proliferative potential. They first grew as clones after limiting dilution and then expanded rapidly with the typical features of spindle-shaped cell bodies and confluence after a lag phase of 6-14 days. There were no differences among the pre and post-cryopreservation of colony formation at the same passage (Fig 2). In addition, the results showed that the passage procedure was selective for MSCs and it could be inferred that the passaging of resuscitated MSCs increased cellular homogeneity.

Ji et al. demonstrated that cryopreservation of encapsulated HES cells offers better cellular viability, higher colony recovery, and less differentiation than the slow-freezing techniques most commonly used to preserve HES colonies. Therefore, this difference in recovery may be due to differences in cell lines, freezing and thawing protocols, or growth substrate (49).

On the other hand, previous studies have shown that cyropreservation had no effect on either the proliferation or osteogenic and adipogenic differentiation of human MSCs *in vitro* (5, 22). In agreement with these reports, Liu et al. using slow cooling with Me2SO as a cryoprotectant and rapid thawing demonstrated that thawed cryopreserved human MSCs had higher survival rates in comparison with non-cryopreserved MSCs and differentiated into osteoblasts when cultured in osteogenic media. Also they found that cryopreserved hMSCs could not differentiate into osteoblasts spontaneously when cultured in basic culture media (50). In addition to the characteristics described above, our present study demonstrated that post-cryopreserved MSCs from bone marrow were still pluripotential and differentiated into osteoblasts and adipocyte under appropriate culture conditions (Fig 3, 4). These observations suggest that the "memory" of proliferation and differentiation in MSCs is not affected by the process of vitrification. The ability of frozen BM-MSCs by slow freezing to differentiate into mesenchymal derivatives (such as osteogenic and adipogenic) reported by Lee et al. (4). In this study, we established a two-step vitrification protocol for MSCs using EFS containing 40% v/v EG for the vitrification solution, which is widely used for successful vitrification of mouse embryos (30), human blastocysts (32) and ESCs (34). Our findings are in line with the reports by Gajda et al. who used the same methods for somatic cells which have been proven to be less toxic on bovine skin fibroblast and cumulus cells (35).

EG is the most commonly used cryoprotectant for vitrification due to its low molecular weight and low toxicity (35). In addition, additives with high molecular weights, such as sucrose, can significantly reduce toxicity by decreasing the concentration of permeating agents required for the vitrification solution. We also used ficoll as a macromolecule to promote permeation by cryoprotectants, which seems to have the advantages of lower toxicity, higher solubility and lower viscosity (30). In a study by Moon et al. on HAMs, they observed that the combination of EG with either PROH or DMSO resulted in a very low survival rate of HAMs as compared with EFS alone (21). Also, Kuleshova and Lopata ascertained the advantages of vitrification when compared with earlier applied cryopreservation techniques (51). These advantages include the control of solute penetration

Cryopreservation of Rat Bone Marrow Derived Mesenchymal Stem Cells by Two Conventional
and Open-Pulled Straw Vitrification Methods

45

and dehydration rates, prevention of prolonged temperature shock and damage from ice formation, and inexpensive equipment and running costs. Vitrification is a process where glass-like solidification of a solution occurs without the formation of ice crystals inside living cells, by exposure to a high concentration of cryoprotectant with a higher cooling rate (52). This procedure is a simple method to circumvent the obstacles of slow freezing without the need for a freezing container to modulate the reduction of temperature in a deep freezer before storing in liquid nitrogen (21).

8. Conclusion

In the present experiment, it was shown that vitrification can be an efficient storage method for MSCs without losing their activity and usual properties. Such a system will be exceedingly helpful for both experimental research and medical applications.

9. Acknowledgments

This research was a part of proposal supported by a grant (3/132/8162 /P) funded by the Guilan University of Medical Sciences and was done in the Cellular and Molecular Research Center at the Faculty of Medicine. The authors wish to thank both centers. There is no conflict of interest in this article.

10. References

[1] Eslaminejad MB, Nazarian H, Taghiyar L. Mesenchymal stem cells isolation from removed medium of rat's bone marrow primary culture and their differentiation into skeletal cell lineage. Yakhteh. 2008; 10(1): 62-72.

[2] Bianco P, Riminucci M, Gronthos S, Robey PG. Bone marrow stromal stem cells: nature, biology, and potential applications. Stem Cells. 2001; 19(3): 180-192.

[3] Wu QY, Li J, Feng ZT, Wang TH. Bone marrow stromal cells of transgenic mice can improve the cognitive ability of an Alzheimer's disease rat model. Neurosci Lett. 2007; 417(3): 281-285.

[4] Lee MW, Choi J, Yang MS, Moon YJ, Park JS, Kim HC, et al. Mesenchymal stem cells from cryopreserved human umbilical cord blood. Biochem Biophys Res Commun. 2004; 320(1): 273-278.

[5] Bruder SP, Jaiswal N, Haynesworth SE, Growth kinetics, self renewal, and the osteogenic potential of purified human mesenchymal stem cells during extensive subcultivation and following cryopreservation. J Cell Biochem. 1997; 64(2): 278-294.

[6] Kang XQ, Zang WJ, Song TS. Xu XL, Yu XJ, Li DL, et al. Rat Bone marrow mesenchymal stem cells differentiation into hepatocytes in vitro. World J Gastroenterol. 2005; 11 (22): 3479-3484.

[7] Jiang Y, Jahagirdar BN, Reinhardt RL, Schwartz RE, Keene CD, Ortiz- Gonzalez XR, et al. Pluripotency of mesenchymal stem cells derived from adult marrow. Nature. 2002; 418: 41-49.

[8] Baharvand H, Mathae KI. Culture condition difference for stablishment of new embryonic stem cell lines from the C57 BL/6 and BALB/c mouse strains. In Vitro Cell Dev Biol Anim. 2004; 40(3-4): 76-81.

[9] Toma C, Pittenger MF, Cahill KS, Byrne BJ, Kessler PD. Human mesenchymal stem cells differentiate to a cardiomyocyte phenotype in the adult murine heart. Circulation. 2002; 105: 93-98.

[10] Kotton DN, Ma BY, Cardoso WV, Sanderson EA, Summer RS, Williams MC, et al. Bone marrow-derived cells as progenitors of lung alveolar epithelium. Development. 2001; 128: 5181-5188.

[11] Tsujigiwa H, Nishizaki K, Teshima T, Takeda Y, Yoshinobu J, Takeuchi A, et al. The engraftment of transplanted bone marrow-derived cells into the olfactory epithelium. Brain Res. 2005; 1052(1): 10-15.

[12] Jeon SJ, Oshima K, Heller S, Edge AS. Bone marrow mesenchymal stem cells are progenitors in vitro for inner ear hair cells. Mol Cell Neurosci. 2007; 34(1): 59-68.

[13] Suzuki H, Taguchi T, Tanaka H, Kataoka H, Li Z, Muramatsu K, et al. Neurospheres induced from bone marrow stromal cells are multipotent for differentiation into neuron, astrocyte, and oligodendrocyte phenotypes. Biochem Biophys Res Commun. 2004; 322(3): 918-922.

[14] Alviano F, Fossati V, Marchionni C, Arpinati M, Bonsi L, Franchina M, et al. Term Amniotic membrane is a high throughout source for multipotent mesenchymal stem cells with the ability to differentiate into endothelial cells in vitro. BMC Dev Biol. 2007; 7:11.

[15] Korbling M, Estrov Z. Adult stem cells for tissue repair. A new therapeutic concept? N Engl J Med. 2003; 349(6): 570-582.

[16] Liechty KW, MacKenzie TC, Shaaban AF, Radu A, Moseley AM, Deans R, et al. Human mesenchymal stem cells engraft and demonstrate site-specific differentiation after in utero transplantation in sheep. Nat Med. 2000; 6(11): 1282-1286.

[17] Xiang Y, Zheng Q, Jia BB, Huang GP, Xu YL, Wang JF, et al. Ex vivo expansion, adipogenesis and neurogenesis of cryopreserved human bone marrow mesenchymal stem cells. Cell Biol Int. 2007; 31(5): 444-450.

[18] Poh M, Boyer M, Solan A, Dahl SL, Pedrotty D, Banik SS, et al. Blood vessels engineered from human cells. Lancet. 2005; 365(9477): 2122-2124.

[19] Guy JA, Wouters R, Successful vitrification of umbilical cord tissue to harvest stem cell for therapies in the future. J Cryobiol/abstract. 2008; 57: 329.

[20] Perry BC, Zhou D, Wu X, Yang FC, Byers MA, Chu TM, et al. Collection, cryopreservation, and characterization of human dental pulp derived mesenchymal stem cells for banking and clinical use. Tissue Eng Part C Methods. 2008; 14(2): 149-156.

[21] Moon JH, Lee JR, Jee BC, Suh CS, Kim SH, Lim HJ, et al. Successful vitrification of human amnionderived mesenchymal stem cells. Hum Reprod 2008; 23(8): 1760-1770.

[22] Kotobuki N, Hirose M, Machida H, Katou Y, Muraki K, Takakura Y, et al. Viability and osteogenic potential of cryopreserved human bone marrow derived mesenchymal cells. Tissue Eng. 2005; 11(5-6): 663-673.

[23] Haack- Sorensen M, Bindslev L, Mortensen S, Friis T, Kastrup J. The influence of freezing and storage on the characteristics and functions of human mesenchymal stromal cells isolated for clinical use. Cytotherapy 2007; 9(4): 328-337.

[24] Lee RH, Kim B, Choi I, Kim H, Choi HS, Suh K, et al. Characterization and expression analysis of mesenchymal stem cells from human bone marrow and adipose tissue. Cell Physiol Biochem. 2004; 14: 311- 324.

[25] Romanov YA, Svintsitskaya VA, Smirnov VN. Searching for alternative sources of postnatal human mesenchymal stem cells: candidate MSC-like cells from umbilical cord. Stem Cells. 2003; 21(1): 105-110.

[26] Zhao J, Hao HN, Thomas RL, Lyman WD. An efficient method for the cryopreservation of fetal human liver hematopoeitic progenitor cells. Stem Cells. 2001; 19(3): 212-218.

[27] Yang PF, Hua TC, Wu J, Chang ZH, Tsung HC, Cao YL. Cryopreservation of human embryonic stem cells: a protocol by programmed cooling. Cryo Letters. 2006; 27(6): 361-368.

[28] Fujioka T, Yasuchika K, Nakamura Y, Nakasuji N, Suemori H. A simple and efficient cryopreservation method for primate embryonic stem cells. Int J Dev Biol. 2004; 48(10): 1149-1154.

[29] Vajta G, Nagy Z. Are programmable freezers still needed in the embryo laboratory? Review on vitrification. Reprod Biomed Online. 2006; 12(6): 779-796.

[30] Kasai M, Mukaida T. Cryopreservation of animal and human embryos by vitrification. Reprod Biomed Online. 2004; 9(2): 164-170.

[31] Karlsson JO. Cryopreservation: freezing and vitrification. Science. 2002; 296(5568): 655-656.

[32] Cho HJ, Son WY, Yoon SH, Lee SW, Lim JH. An improved protocol for dilution of cryoprotectants from vitrified human blastocysts. Hum Reprod. 2002;17(9): 2419-2422.

[33] Reubinoff BE, Pera MF, Vajta G, Trounson AO. Effective cryopreservation of human embryonic stem cells by the open pulled straw vitrification method. Hum Reprod. 2001; 16(10): 2187-2194.

[34] Nie Y, Bergendahl V, Hei DJ, Jones JM, Palecek SP. Scalable culture and cryopreservation of human embryonic stem cells on microcarriers. Biotechnol Prog. 2009; 25(1): 20-31.

[35] Gajda B, Katska- Ksiazkiewicz L, Rynska B, Bochenek M, Smorag Z. Survival of bovine fibroblasts and cumulus cells after vitrification. Cryo letters. 2007; 28(4): 271-279.

[36] Stock P, Staege MS, Muller LP, Sgodda M, Volker A, Volkmer I, et al. Hepatocyte Derived from adult stem cells. Transplant Proc. 2008; 40(2): 620-623.

[37] Polisetty N, Fatima A, Madhira SL, Sangwan VS, Vemuganti GK. Mesenchymal cells from limbal stroma of human eye. Mol Vis. 2008; 14: 431-442.

[38] Wang QR, Wolf NS. Dissecting the hematopoietic microenvironment. Clonal isolation and identification of cell types in murine CFU-F colonies by limiting dilution. Exp Hematol. 1990; 18(4): 355-359.

[39] Eslaminejad MB, Nikmahzar A, Taghiyar L, Nadri S, Massumi M. Murine mesenchymal stem cells isolated by low density primary culture system. Dev Growth differ. 2006; 48(6): 361-370.

[40] Vajta G, Kuwayama M. Improving cryopreservation systems. Theriogenology. 2006; 65: 236–244.

[41] Wang HY, Lun ZR, Lu SS. Cryopreservation of umbilical cord blood-derived mesenchymal stem cells without dimethyl sulfoxide. Cryo Letters. 2011; 32(1): 81-88.

[42] Kim GA, Lee ST, Lee EJ, Choi JK, Lim JM. Simplified slow freezing program established for effective banking of embryonic stem cells. Asian-Aust J Anim Sci. 2009; 22(3): 343-349.

[43] Katkov MS, Kim R, Bajpai YS, Altman M, Mercola JF, Loring AV, et al. Cryopreservation by slow cooling with DMSO diminished production of Oct-4 pluripotency marker in human embryonic stem cells. Cryobiol. 2006; 53:194-205.

[44] Suemori H, Yasuchika K, Hasegawa K, Fujioka T, Tsuneyoshi N, Nakatsuji N. Efficient establishment of human embryonic stem cell lines and long-term maintenance with stable karyotype by enzymatic bulk passage. Biochem Biophys Res Commun. 2006; 345:926-932.

[45] Carvalho KAT, Cury CC, Oliveira L, Cattaned RII, Malvezzi M, Francisco JC, et al. Evaluation of Bone Marrow Mesenchymal Stem Cell Standard Cryopreservation Procedure Efficiency. Transplant Proc. 2008; 40: 839-841.

[46] Yang PF, Hua TC, Tsung HC, Cheng QK, Cao YL. Effective cryopreservation of human embryonic stem cells by programmed freezing. Conf Proc IEEE Eng Med Biol Soc. 2005; 1: 482-485.

[47] Katkov II, Kan NG, Cimadamore F, Nelson B, Snyder EY, Terskikh AV. DMSO-Free Programmed Cryopreservation of Fully Dissociated and Adherent Human Induced Pluripotent Stem Cells. Stem Cells Int. 2011;2011:981606.

[48] Heng BC. Effect of Rho-associated kinase (ROCK) inhibitor Y-27632 on the post-thaw viability of cryopreserved human bone marrow- derived mesenchymal stem cells. Tissue Cell. 2009;41(5):376-380.

[49] Ji L, de Pablo JJ, Palecek SP. Cryopreservation of adherent human embryonic stem cells. Biotechnol Bioeng. 2004; 88(3): 299-312.

[50] Liu G, Shu C, Cui L, Liu W, Cao Y. Tissue-engineered bone formation with cryopreserved human bone marrow mesenchymal stem cells. Cryobiology. 2008; 56(3): 209-215.

[51] Kuleshova L, Lopata A. Vitrification can be more favorable them slow cooling. Fertil Steril. 2002; 78: 449-454.

[52] Chian RC, Kuwayama M, Tan L, Tan J, Kato O, Nagai T. High survival rate of bovine oocytes matured in vitro following vitrification. J Reprod Dev. 2004; 50: 685-696.

Cryopreservation of Adherent Smooth Muscle and Endothelial Cells with Disaccharides

Lia H. Campbell[1] and Kelvin G.M. Brockbank[1,2,3]
[1]Cell & Tissue Systems, Inc., North Charleston, SC
[2]Georgia Tech / Emory Center for the Engineering of Living Tissues, The Parker H. Petit
Institute for Bioengineering and Bioscience, Georgia Institute of Technology, Atlanta, GA
[3]Medical University of South Carolina, Department of Regenerative Medicine and Cell
Biology, Charleston, SC
USA

1. Introduction

There is a need for mammalian cell cryopreservation methods that either avoid or improve upon outcomes employing dimethyl sulfoxide (DMSO) as a cryoprotectant. DMSO was the second effective cryoprotectant to be discovered (Lovelock, 1959). Cell cryopreservation usually involves slow rate freezing with DMSO in culture medium and storage below -135°C for later use. Typically as long as there are enough cells surviving to start an expanding proliferating culture the yield of viable cells after thawing is not an important consideration. However, there are instances where cell yield and viability can be very important. Examples include minimization of expensive delays when starting cultures for bioreactor protein manufacturing runs and cellular therapies that involve administering cells into patients for treatment of various diseases, such as cancer. While some cells, for example fibroblasts, are easily cryopreserved other cell types like keratinocytes, hepatocytes, and cardiac myocytes do not freeze well and cell yields are often <50%. Furthermore, current opinion is that DMSO should be removed before cells are infused into patients (Caselli et al., 2009; Junior et al., 2008; Mueller et al., 2007; Otrock et al., 2008; Schlegel et al., 2009). The mechanism for DMSO cytotoxicity has not been determined, however, it is thought to modify membrane fluidity, induce cell differentiation, cause cytoplasmic microtubule changes and metal complexes (Barnett 1978; Katsuda et al., 1984, 1987; Miranda et al., 1978). DMSO also decreases expression of collagen mRNAs in a dose-dependent manner (Zeng et al., 2010).

One strategy for finding interesting new cryoprotectants and cryopreservation strategies is by evaluating what happens in nature (Brockbank et al., 2011). No examples of organisms synthesizing DMSO to survive freezing conditions have been found to date, however several creatures have been found that employ glycerol (Brockbank et al., 2011) the first effective cryoprotectant to be discovered (Polge, 1949). Nature has developed a wide variety of organisms and animals that tolerate low temperatures and dehydration stress by accumulation of large amounts of disaccharides, particularly trehalose, including plant seeds, bacteria, insects, yeast, brine shrimp, fungi and their spores, cysts of certain crustaceans, and some soil-dwelling animals. While the cryoprotective capabilities of sucrose and trehalose has been known for years, conventional cryopreservation protocols have generally not employed them even though early work with them demonstrated their ability to protect proteins and membrane vesicles during freezing (Rudolf & Crowe, 1985; Crowe et al., 1990). Trehalose has both major advantages and disadvantages for potential preservation of mammalian cells. On the negative side mammalian cells do not have an active trehalose transport system for uptake of trehalose from the extracellular environment, while on the plus side if you can get it in mammalian cells it is not metabolized giving the opportunity for trehalose to be accumulated to potentially effective preservation concentrations. The purpose of the studies presented here were; 1) to assess or review alternative strategies for delivery of trehalose into mammalian cells, and; 2) to determine whether the benefits were specific to trehalose by investigating alternative sugars employing the same loading strategies.

2. Materials and methods

2.1 Cell culture

Cells used in these studies are described in Table 1. All were grown and maintained at 37°C in 5% carbon dioxide.

Description	Acronym	Culture conditions
Rat aortic myofibroblast cells	A10 (ATCC# CRL-1476)	DMEM* (4.5 g/L) 10% FCS 1.0 mM sodium pyruvate
Bovine calf pulmonary artery endothelial cells	CPAE (ATCC# CCL-209)	EMEM** 10% FCS 1 mM sodium pyruvate 2 mM glutamine 1X non-essential amino acids
Rat aortic smooth muscle cells	A7R5 (ATCC# CRL-1444)	DMEM (4.5 g/L) 10% FCS 1.0 mM sodium pyruvate
Bovine corneal endothelial cells	BCE (ATCC# CRL-2048)	DMEM (4.5 g/L) 10% FCS 1 mM sodium pyruvate 4 mM glutamine

*Dulbecco's Modified Eagle's Medium
**Eagle's Minimum Essential Medium

Table 1. Cell types

2.2 Cell poration with H5

The pore-forming protein H5 was obtained from the lab of Hagan Bayley (Bayley, 1994). It is derived from the bacterial toxin α-hemolysin, which forms constitutively opened pores in cell membranes. The modified bacterial toxin has been engineered to form pores in the membrane that can be opened and closed by the addition of Zn^+. More specific details are presented in the discussion. Cells were plated at 10,000-20,000 cells/well the night before in 96 well microtiter plates. The next day, the cells were washed with DMEM containing 1 mM EDTA for 2 minutes and then again with DMEM to remove the EDTA. 0.2M trehalose was added and incubated for 20 minutes at 37°C followed by the appropriate concentration of H5 for the respective cell type. Cells were porated and loaded with trehalose for 1 hour at 37°C before addition of DMEM with 25 μM $ZnSO_4$ or 10% serum to close the pores. Trehalose in DMEM was then added to the wells followed by cryopreservation using a controlled rate freezer (Planar) at ~-1.0°C/min from 4°C to -80°C with a programmed nucleation step at -5.0°C. Cryopreserved cells were stored overnight at <-135°C. The next day, the cells were placed at -20°C for 30 minutes followed by rapid thawing at 37°C (Campbell et al., 2003; Taylor et al., 2001). The cell cultures were washed twice and then placed at 37°C for 1 hour to recover under normothermic cell culture conditions before assessment of cell viability.

2.3 Pretreatment (Incubation) with trehalose

Cells were plated at 10,000-20,000 cells/well and placed in culture. The next day, the culture medium was replaced with EMEM or DMEM containing trehalose (0-0.6M) and cultured at 37°C for varying periods of time. After culture, the solution was replaced with fresh medium containing trehalose (0-0.6M) and the cells were cryopreserved using a controlled rate freezer as described for H5 above.

2.4 Cell poration with ATP

Cells were plated at 10,000-20,000 cells/well and placed in culture. The next day, the cells were washed with poration buffer (phosphate-buffered saline [PBS] with 1X essential amino acids, 1X Vitastock, 5.5 mM glucose) designed to optimize binding of ATP^{4-} to the receptor and facilitate formation of the pore. The cells were then placed in 50 μl poration buffer, pH of 7.45, with 0.2M trehalose. A stock solution of 100 mM ATP^{4-}, pH of 7.45, was made fresh and added to each well to achieve a final concentration of 5 mM. After addition of the ATP^{4-}, the cells were left at 37°C for 1 hour to allow sugar uptake. Following incubation, 200 μl of DMEM plus 1 mM $MgCl_2$ was then added to the cells at 37°C to close the pores. After 1 hour of recovery from the loading procedure cryopreservation was initiated.

2.5 Assessment of cell viability

Cell viability was determined using the non-invasive metabolic indicator alamarBlue™ (Trek Diagnostics). A volume of 20 μl was added to cells in 200 μl of DMEM (10%FCS) and the plate was incubated at 37°C for 3 hours. Plates were read using a fluorescent microplate reader (Molecular Dynamics) at an excitation wavelength of 544 nm and an emission wavelength of 590 nm. Viability was measured before and after sugar loading, immediately after thawing and at several later time points post-thaw.

2.6 Statistical methods

All experiments were repeated at least four times with four replicates in each experiment. Statistical differences were assessed by two way analysis of variance. P-values <0.05 were regarded as significant.

3. Results

3.1 H5 poration

One of the first strategies for utilizing disaccharide sugars as cryoprotectants involved the use of a modified pore forming complex. In our initial studies we evaluated the H5 mutant α-hemolysin (Bayley, 1994) using two adherent cell lines, A10 and CPAE. The earlier studies had been done with cells in suspension (Eroglu et al., 2000). We also evaluated sucrose, another disaccharide sugar that is commonly found in nature, for its potential usefulness as a cryoprotectant. Using the protocol of Eroglu et al as a starting point, a protocol was established for adherent cells. Several parameters were evaluated and included the H5 concentration, time of poration, concentration of trehalose loaded, and time for loading trehalose. Conditions that worked best with adherent cells included 20 minutes for poration followed by 60 minutes for trehalose loading. The highest concentration of trehalose that caused the least drop in cell viability was 0.2M. The optimum H5 concentration varied according to cell type. The A10 smooth muscle cells were porated with 12.5 μg/mL of H5 while the endothelial CPAE cells were porated with 50 μg/mL. In contrast, the fibroblasts and keratinocytes in the literature were porated with 25 μg/mL (Eroglu et al., 2000). Other changes to the protocol were made that benefited viability for adherent cells specifically and included addition of trehalose prior to the addition of H5, the base solution used for poration, and the amount of EDTA (1 mM versus 10 mM) for the removal of Zn^+ prior to poration. After cryopreservation, however, poor viability was obtained with both cell types. A10 cells demonstrated a viability of 5.57±0.17%. The endothelial cells demonstrated similar viabilities. These values were not as good as those observed when suspended cells were cryopreserved with sugars in the literature. However, it is our experience that adherent cells are generally more difficult to cryopreserve regardless of the cryoprotectant used.

3.2 Trehalose exposure without poration

When we started adding the trehalose to cells in the H5 experiments, control cells were exposed to trehalose by addition to the culture medium prior to cryopreservation. An unanticipated observation of cell survival was made with slow rate cryopreserved CPAE cells prompting further investigation. The cells exposed to trehalose overnight were observed to develop vacuoles (Fig 1) suggesting a possible pinocytotic uptake mechanism.

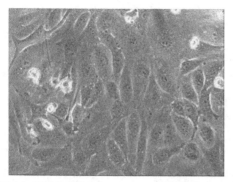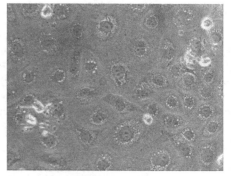

Fig. 1. CPAE cells after exposure to trehalose. Left: no sugar, Right: 0.1M trehalose. 40X magnification

After these observations were made, further experiments were designed to examine cell viability after extended trehalose exposure. CPAE cells were exposed to 0.2M trehalose in Dulbecco's Modified Eagle's Medium (DMEM) buffered with 25 mM Hepes for 0-72 hours at 37°C. After exposure the cells were left in 0.2M trehalose and cryopreserved at ~-1.0°C/min (Fig. 2). CPAE cell viability was observed immediately after thawing. An exposure time of 24 hours provided the best overall cell survival. Extracellular exposure alone during cryopreservation failed to produce any cell survival. In contrast, A10 smooth muscle cells generally did not survive cryopreservation after trehalose exposure as well as the CPAE endothelial cells. Examination of optimal concentrations of trehalose during incubation and during cryopreservation showed that a concentration of 0.1-0.2M trehalose for incubation produced the best viability with a similar concentration being required during the freezing process.

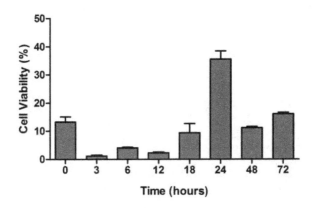

Fig. 2. Impact of cell culture time with trehalose on cryopreservation. CPAE cells were cultured with 0.2M trehalose for up to 72 hours followed by cryopreservation with 0.2M trehalose. Percent viability was calculated based on the pre-cryopreservation controls. (p<0.05)

Several other parameters were also examined to further improve cell viability. Other studies have shown that not only the concentration and choice of cryoprotectant but also the vehicle solution for the cryoprotectant can have a significant impact on cell viability after cryopreservation (Campbell & Brockbank, 2007; Mathew et al., 2004; Sosef et al., 2005). Initial experiments were performed using DMEM, however, it was observed that CPAE cells, which are grown in EMEM medium, actually preferred exposure to trehalose in EMEM medium. Further experiments examined the buffers used to maintain the pH of the system. Four cell lines were evaluated. CPAE cells demonstrated decreased viability when the zwitteronic buffer, 4-(2-hydroxyethyl)-1-piperazineethanesulfonic acid (HEPES) was used while the other 3 cell lines did not show decreased viability. Rather a combination of HEPES and sodium bicarbonate was preferred by the CPAE cells. This unusual choice of buffer prompted examination of solution pH during incubation, a pH of 7.4 was optimal for all the 4 cell lines tested. Once loaded with sugar, the cells could either be left in the extracellular sugar at another concentration or an alternative cryoprotectant for preservation.

These studies were then extended to include other sugars, sucrose, raffinose, and stachyose (Fig. 3). The potential cryoprotective benefits of these sugars were evaluated and it was found that stachyose was as good as trehalose using an identical protocol, sucrose was not quite as good and raffinose had very little benefit. All cell lines showed evidence of some cell survival days after cryopreservation and thawing. The second smooth muscle cell line, A7R5, demonstrated low levels of viability with stachyose. Both endothelial cell lines, CPAE and BCE, showed good viability after exposure and freezing with sucrose. Overall, the CPAE cell line had the best viability in these experiments. Use of an optimized protocol with trehalose produced excellent post cryopreservation results with 10-14mM intracellular trehalose (Campbell, 2011). Conditions included 24 hours of cell culture with 0.2M trehalose followed by cryopreservation with 0.2-0.4M trehalose in sodium bicarbonate buffered EMEM at pH 7.4 resulting in ~75% post-preservation cell viability (Campbell et al., 2011).These experiments confirmed that this technique is more effective for endothelial cells than smooth muscle cells and demonstrated that stachyose is effective for cryopreservation.

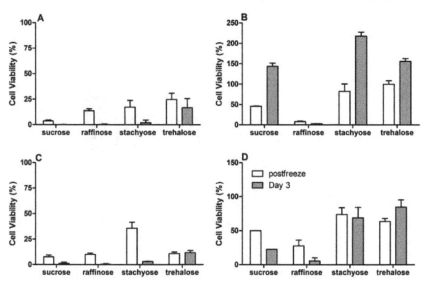

Fig. 3. Cell viability for A10 (A), CPAE (B), A7R5 (C) and BCE (D) cells after exposure and freezing in the presence of sugars.

3.3 ATP poration

In addition to the H5 mutant α-hemolysin poration strategy, we sought other poration techniques that could be used to permeate mammalian cells with disaccharides. Cells expressing the $P2_{X7}$ purinergic cell surface receptor, also known as the $P2_z$ receptor, may be permeabilized by the formation of a channel/pore that allows passage of molecules into and out of the cell when the active form of ATP (ATP^{4-}) binds to the receptor. Our initial studies focused on determination of whether or not the $P2_{X7}$ was expressed on smooth muscle and endothelial cells. Experiments using the ELICA assay demonstrated the presence of the $P2_{X7}$ receptor on both endothelial and smooth muscle cell lines to varying degrees (Fig 4). The smooth muscle cell lines demonstrated the greatest density of the receptor.

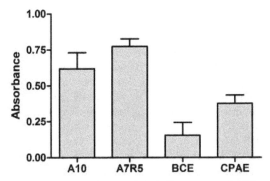

Fig. 4. Detection of the P2$_{X7}$ receptor by ELICA. Cells were probed for the presence of the P2$_{X7}$ receptor using antiserum specific for the receptor at a dilution of 1:25. The graph represents the average absorbance (±SEM) for 10 replicates.

ATP-permeabilized cells retained better viability than untreated cells both immediately after thawing and five days later (Fig 5). Immediate metabolic activity in A7R5 and CPAE cells demonstrated dependence upon increasing ATP concentrations, while for A10 and BCE cells immediate metabolic activity was increased at all ATP concentrations with only slight improvement at the higher concentrations tested. However, survival at five days demonstrated that intermediate concentrations of ATP (0.5-2.5mM) were best. Further cryopreservation studies were performed to optimize cell survival resulting in at least 25% cell survival for both endothelial cell lines but only low levels of survival for the smooth muscle cells.

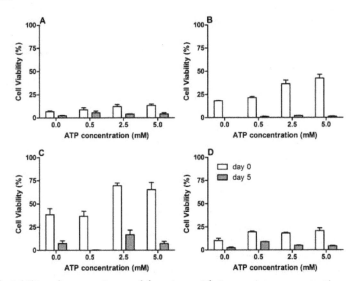

Fig. 5. Cell viability after poration and freezing with increasing concentrations of ATP. Cells were loaded with 0.2M trehalose using the P2$_{X7}$ receptor and the indicated concentrations of ATP. After poration and cryopreservation, cell viability was evaluated by alamarBlue. (A) A10, (B) A7R5, (C) CPAE, (D) BCE.

4. Discussion

As cryopreservation has been applied to cells and tissues for clinical use, concerns about toxicity relating to the various cryoprotectants being used, particularly DMSO, have developed. Because of this, there has been renewed interest in finding less toxic cryoprotectants. The cryoprotective capabilities of some sugars, disaccharide sugars in particular, has been known for years and early work with them demonstrated their ability to protect proteins and membrane vesicles during freezing (Crowe et al., 1990; Rudolph & Crowe, 1985). Coupled with these early studies are observations made in nature regarding organisms that can survive extremes in temperature and desiccation due to their ability to accumulate large amounts of disaccharide sugars, specifically trehalose and sucrose, until more favorable conditions are available. The protective effects of trehalose and sucrose have been determined and may be classified under two general mechanisms: (1) "the water replacement hypothesis" or stabilization of biological membranes and proteins by direct interaction of sugars with polar residues through hydrogen bonding, and (2) stable glass formation (vitrification) by sugars in the dry state (Crowe et al., 1987, 1988, 1998, 2001; Slade & Levine, 1991).

Two primary stresses that destabilize membranes have been defined, fusion and lipid phase transition. Studies have shown that when the water that hydrates the phospholipid molecules of the membrane is removed, packing of the head groups increases. The result is an increase in van der Waals interactions and a dramatic increase in the phase transition temperature (T_m) (Crowe et al., 1987, 1988, 1990, 1991). At the phase transition the phospholipid bilayer shifts from a gel phase to a liquid crystalline phase, the state normally observed in fully hydrated cells. For example, the T_m of a cell membrane might be -10°C when fully hydrated but when water is removed the T_m increases to over 100°C. Thus, the membrane is in the gel phase at room temperature. As the membrane shifts between the gel phase and the liquid crystalline phase it becomes transiently leaky allowing its intracellular contents to leak out. Therefore it would be advantageous to avoid the lipid phase transition as this can compromise the health of a rehydrated cell. Addition of disaccharide sugars, in particular trehalose, depresses T_m allowing the membrane to remain in the liquid crystalline state even when dried, so that upon rehydration no phase transition takes place and no transient leaking. During cryopreservation water is not necessarily lost, but it undergoes a phase change forming ice as the temperature drops and depending upon the rate of cooling, the cells become more or less dehydrated rendering the cells vulnerable to damage by mechanisms similar to those proposed for desiccated cells.

The stabilizing effect of these sugars has been shown in a number of model systems including liposomes, membranes, viral particles, and proteins. The mechanism by which disaccharide sugars are able to decrease the T_m for a given bilayer has been elucidated. Interactions take place between the sugars and the –OH groups of the phosphate in the phospholipid membrane preventing interaction or fusion of the head groups as the structural water is removed (Crowe et al., 1986, 1988, 1989a, 1989b). Although not as well understood, a similar mechanism of action stabilizes proteins during drying (Carpenter et al., 1986, 1987a, 1987b, 1989). Despite their protective qualities, the use of these sugars in mammalian cells has been somewhat limited mainly because mammalian cell membranes are impermeable to disaccharides or larger sugars and there is strong evidence that sugars need to be present on both sides of the cell membrane in order to be effective (Crowe et al.,

2001; Eroglu et al., 2000; Beattie et al., 1997). This is why, in addition to loading sugars, we added sugars to the cryopreservation solution just before initiating cooling.

In addition to trehalose and sucrose, we were interested in other sugars that could be used as cryoprotectants avoiding monosacharides that would likely be degraded in the cell. Larger more complex sugars such as disaccharides or larger would be less likely to be degraded and utilized inside cells and might therefore be more stable as cryoprotectants. The comparative structures of the sugars we considered for preservation of mammalian cells are illustrated in Figure 6. Three other sugars were evaluated besides trehalose and included sucrose, raffinose and stachyose. Sucrose and trehalose are both non-reducing sugars, so they do not react with amino acids or proteins and should be relatively stable under low pH conditions and at temperature extremes. Raffinose is a trisaccharide and stachyose is a tetrasaccharide.

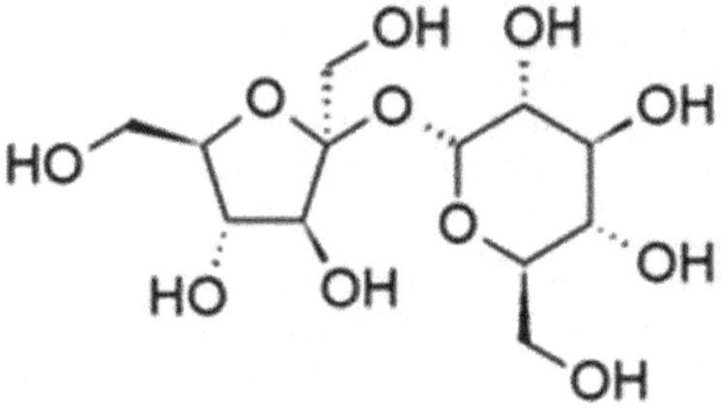

Sucrose (mw: 342.30) 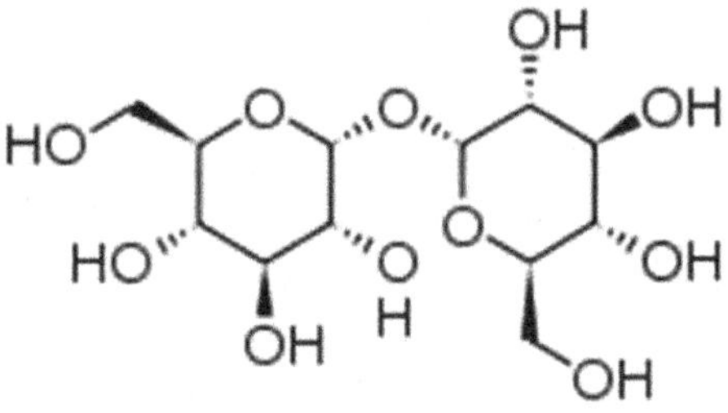   Trehalose (mw: 342.30)

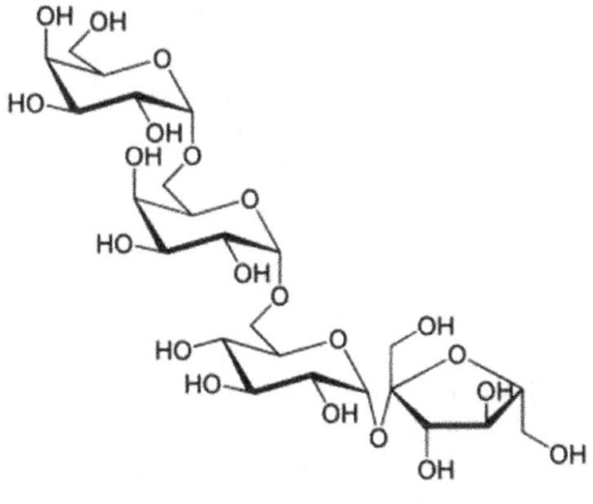

Stachyose (mw: 684.59) 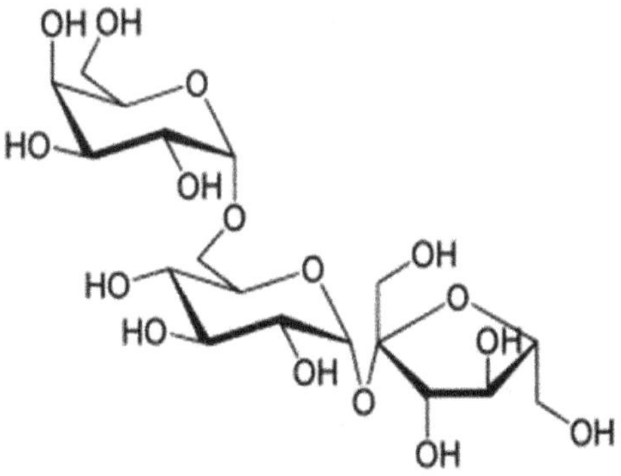   Raffinose (mw: 504.44)

Fig. 6. Sugar structure

Before going further, it is important to point out that the cells we have employed were cryopreserved and thawed while adherent in 96-well plates using cooling and warming conditions defined and reported at the turn of the century (Campbell et al., 2003; Taylor et al., 2001). We have since used these conditions to cryopreserve several adherent cell types (Campbell et al., 2007, 2010, 2011). Our rationale for using this adherent model was two-fold. First, due to our interest in regenerative medicine we thought that adherent cells more closely mimicked cells in tissue engineered devices. Second, we thought there might be a market for cells cryopreserved on plates for research and cytotoxicity testing, CryoPlate™. More recently another group has been using adherent cells for investigation of preservation by vitrification and drying and have reported on cryopreservation of adherent pluripotent stem cells (Katkov et al., 2006; Katkov et al., 2011;). Katkov et al. presented results for preservation of human embryonic stem cells in 4-well plates and pointed out several advantages of cryopreservation in adherent mode. These included elimination of possible bias due to selective pressure within a pluripotent stem cell line after cryopreservation and distribution of multiwell plates for immediate use for embryotoxicity and drug screening in pluripotent stem cell-based toxicology in vitro kits (Katkov et al., 2011).

There are several methods in the literature that could be employed for intracellular delivery of these sugars including those already discussed (Campbell et al., 2011; Table 2). Many drugs, therapeutic proteins and small molecules have unfavorable pharmacokinetic properties and do not readily cross cell membranes or other natural physiological barriers within the body. This has resulted in the search for and discovery of alternative methods to transport materials, like sugars, across mammalian cell membranes.

Some of these strategies have been presented in depth in the results sections. The first involved the use of the *Staphylococcus aureus* toxin, α-hemolysin. This toxin is produced as a monomer by the bacteria. It then oligomerizes to form pores on mammalian cell membranes. Hagan Bayley and his group modified the wild type α-hemolysin protein by replacing 4 native residues with histidines, termed H5. In addition to pore formation on cell membranes, the H5 mutant also enabled it to be opened and closed at will. When inserted into the membrane, it is open and molecules up to 3000 daltons are able to pass through. Then the pores are closed in the presence of Zn^+. To reopen the pore, addition of a chelating agent such as EDTA will remove the Zn^+ and the pore is ready to be used again (Bayley, 1994; Walker et al., 1995). Early studies showed that H5 could create pores in mammalian cell membranes and that they could be used for efficient intracellular loading of trehalose (Eroglu et al, 2000; Acker et al., 2003). Our experiments with H5 worked well initially using adherent cells. The results demonstrated good poration and loading of trehalose into cells. However, after adherent cells were cryopreserved, their viability was not very good (<6%). At this point in our studies, several issues arose that prevented further studies using H5. First, the H5 pore was derived from the bacterial toxin α-hemolysin so there were concerns raised whether regulatory approval could be obtained if it was ever to be used clinically with human cells and tissues. There were some indications during these studies that the pores were shed from the membrane over time. However, H5 was still detectable in picogram quantities after 7 days in culture. Finally, as new batches of H5 were delivered the activity varied greatly and more H5 was required to achieve the same level of poration compared with earlier batches. Ultimately the batch variation was attributed to a protein stability issue. When these issues were not resolved other strategies for introducing trehalose into cells were explored.

An unexpected outcome of our H5 experiments was the development of a new, simple strategy for introduction of trehalose into cells which involved incubating cells in sugar for extended periods of time at physiological temperature (Brockbank, 2007). One possible mechanism to explain this observation was that the trehalose is substituting for water molecules in the cell membranes keeping the membrane stable and preventing it from going through a phase transition (Crowe et al., 1988, 1989). A second mechanism is most likely an active uptake mechanism involving endocytosis similar to that proposed for loading of trehalose by Oliver et al (Oliver et al., 2004). Her results suggested that human MSCs are capable of loading trehalose from the extracellular space by a clathrin-dependent fluid-phase endocytotic mechanism that is microtubule-dependent but actin-independent (Oliver et al., 2004). Further research is required to elucidate the mechanism by which culture in the presence of trehalose facilitates cell cryopreservation and determine the degree of cell viability retention under different storage conditions.

The last method presented was poration using the $P2_z$ receptor and ATP. This was a somewhat unique strategy in that it took advantage of the cell's own machinery. It was shown that cells expressing the $P2_{X7}$ purinergic cell surface receptor, also known as the $P2_z$ receptor, could be permeabilized when the receptor binds to ATP^{4-}. The interaction with ATP resulted in the formation of a non-selective pore that allows molecules up to ~900 Daltons to pass through (Nihei et al., 2000). The $P2_{X7}$ receptor selectively binds to only ATP^{4-} whose presence in solution is dependent on temperature, pH, and the concentration of divalent cations such as Mg^{2+}. Closure of the pore after activation by ATP is achieved by simply removing ATP from the system or adding exogenous Mg^{2+} that has a high affinity for the active form of ATP, ATP^{4-}. The $P2_z$ receptor is found on a number of cell types including cells of hematopoietic origin (Nihei et al, 2000). There were several factors that likely affected the cell viability and survival of cells after ATP poration. First, is the density of the receptor on the cells which directly affects the amount of trehalose that can be loaded into the cells and how long it takes. Another factor is that poration with ATP tends to promote the detachment of adherent cells from their substrate. Part of the protocol requires a recovery period of 1 hour at 37°C to allow cells that may have been perturbed by the poration process the chance to settle back onto their substrate. Finally, cell loss is at least in part due to apoptosis. There is evidence in the literature that poration with ATP induces apoptosis in some cell types (Murgia et al., 1992).

In marked contrast the human stem cell line, TF-1, demonstrated excellent post cryopreservation survival (Buchanan et al., 2004; 2005). We have exposed TF-1 cells to ATP with trehalose for 1 hour followed by a 10-fold dilution of the ATP and inactivation of the active form of ATP (ATP $^{4-}$) by the addition of 1 mM $MgCl_2$ followed by a 1-hour recovery period at 37°C (Brockbank et al., 2011). When the cells were compared to cells cryopreserved with 10% DMSO, the DMSO group demonstrated greater initial viability close to 100% that steadily declined over days in culture post thaw. However by day 4 of culture post-cryopreservation cells cryopreserved in disaccharides were similar to the viability of cells cryopreserved in DMSO. Similarly colony forming assays with TF-1 cells demonstrated similar outcomes compared with DMSO. Furthermore, the use of disaccharides, trehalose and sucrose, appeared to result in similar results at both slow (1°C/min) and rapid (100°C/min) cooling rates. Buchanan et al (Buchanan et al., 2010) have extended these studies obtaining excellent TF-1 cell line and cord blood-derived multipotential hematopoietic progenitor cell survival after freeze drying and storage at room temperature for 4 weeks! It is studies such as Buchanan's that keep us optimistic that disaccharide introduction/preservation strategies can be developed for preservation of other mammalian cell types. Further development work is required with the cell culture and $P2_{x7}$ methods with the promise of preservation by freezing and freeze-drying.

Existing Techniques	Description	Pitfalls	References
H5	Derived from α-hemolysin, which normally forms a constitutively opened pore in the membrane. Engineered to close in the presence of Zn+ or serum.	Derived from a bacterial toxin. Batch to batch variation and instability.	Acker et al. 2003 Bayley, 1994 Eroglu et al. 2000
ATP	The naturally occurring $p2_{x7}$ receptor forms a non-specific pore upon binding of ATP^{4-} able to allow molecules <900 daltons to pass through.	$P2_{x7}$ receptor found on some but not all cell types.	Buchanan et al. 2005
Culture methods	1) Prolonged incubation of cells in the presence of disaccharide sugars at 37°C. 2) Fluid phase endocytosis: disaccharide sugars are taken up by cells via a clathrin dependent endocytotic mechanism.	Works better with some cells but not others.	Brockbank et al. 2007 Oliver et al. 2004
Temperature manipulation	A shift in temperature can cause a lipid phase transition which temporarily changes the membrane permeability and allows molecules to pass through.	Has been demonstrated with pancreatic islets and kidney cells. Requires optimization by cell type.	Beattie et al. 1997 Mondal 2009

Table 2. Strategies for Loading Disaccharide Sugars

There are still other methods in the literature that could lead to intracellular delivery of disaccharides in addition to those already discussed (Campbell et al., 2011; Table 2). One method takes advantage of the lipid phase transition described above when the cell membrane is exposed to changes in temperature. As the membrane changes from the liquid crystalline phase to the gel phase it becomes leaky providing an opportunity to introduce molecules into the cell that would not normally cross like trehalose. Beattie used this method to cryopreserve pancreatic islets by introducing DMSO and trehalose into the islets during the thermotropic phase transition between 5 and 9°C. The islets were then cryopreserved in combination with DMSO and the viability of the islets after thawing was greater than when DMSO alone was used, 94% versus 58% (Beattie et al., 1997). In a related study, Mondal et al, cryopreserved kidney cells (MDBK) using 264 mM trehalose. The cells were suspended in trehalose with 20% fetal bovine serum in culture medium then incubated at 40°C for 1 hour before slow rate cooling for storage at -80°C. Viability was measured using Trypan Blue exclusion at 74% upon thawing (Mondal, 2009).

In another variation for loading molecules into cells, a number of proteins have been discovered that possess the ability to cross the cell membrane. These protein transduction domains (PTDs) generally correspond to portions of native proteins. Examples of PTDs include the Tat protein from the human immunodeficiency virus type I, the envelope glycoprotein E^{rns} from the pestivirus and the DNA binding domains of leucine zipper proteins such as c-fos, c-jun and yeast transcription factor GCN4 (Futaki et al., 2001, 2004; Langedijk, 2002; Langedijk et al., 2004; Lindgren et al., 2000; Richard et al., 2003; Vives et al., 1997). These PTDs are short cationic peptides that cross the cell membrane in a concentration-dependent manner that is independent of specific receptors or other transporters. The exact mechanism of translocation has not been defined. Enrichment of basic amino acids, particularly arginine and in some instances lysine, have been shown to be important for the translocation activity (Futaki et al., 2001, 2004; Vives et al., 1997). Some studies have suggested that endocytosis is involved (Lundberg & Johansson, 2002; Richard et al., 2003), however, the current theory includes interaction with glycosaminoglycans and uptake by a non-endocytotic mechanism that may involve the charged heads of the phospholipid groups within the cell membrane. (Langedijk, 2002; Langedijk et al., 2004; Mai et al., 2002).

While most of these peptides need to be cross linked to the molecule of interest, there are peptides that can move proteins and other peptides across the membrane without the requirement for cross-linking. Examples include Pep-1, a 21-residue peptide which contains three domains; a tryptophan rich region (5 residues) for targeting the membrane and forming hydrophobic interactions; a lysine rich domain to improve intracellular delivery whose design was taken from other nuclear localization sequences from other proteins like the simian virus 40 large T antigen, and, a spacer region with proline that provides flexibility and maintenance of the other two regions. When mixed with other peptides or proteins, Pep-1 rapidly associates and forms a complex with the protein of interest by noncovalent hydrophobic interactions to form a stable complex. Once in the cytoplasm the peptide dissociates from the protein that has been carried across the membrane causing little if any interference regarding the protein's final destination or function. The process occurs by an endocytosis independent mechanism (Morris et al., 1999, 2001). We anticipate that such peptides may eventually lead to methods for introduction of disaccharides into mammalian cells (Campbell et al., 2011).

Another alternative method is electroporation, also called electropermeabilization, which involves the application of an electric pulse that briefly permeabilizes the cell membrane. Since its introduction in the 1980's it has been primarily used to transfect mammalian cells and bacteria with genetic material. Initially electroporation tended to kill most cells. However, further work and development of the electroporation process, such as alternate electrical pulses like the square wave pulse, have refined the process so that better permeabilization and cell viability can be achieved (Gehl, 2003; Hapala, 1997; Heiser, 2000). The formation of pores, their size and the recovery of the membrane are important factors that influence the success of an electroporation protocol (Gehl, 2003; Hapala, 1997; Heiser, 2000). Most importantly, electroporation is applicable to all cell types.

It was hypothesized that trehalose provided protection during electropermeabilization in a manner similar to chelating agents such as EDTA or lipids like cholesterol (Katkov, 2002;

Mussauer et al., 2001). Effective electroporation protocols are a balance between how much material can be loaded into the cells and cell survival after membrane permeabilization. So, while it cannot be predicted how well certain cell types will respond to electroporation, there is ample evidence that electroporation can be used with a reasonably certainty of success. A short culture period may be all that is required to permit restabilization of membranes post-electroporation. Additionally, like trehalose which interacts with membranes under stressful conditions such as drying, other compounds, such as cholesterol and unsaturated fatty acids, can also interact with membranes and may facilitate resealing of the membranes increasing overall cell survival (Katkov, 2002). Efficient resealing of cell membranes after permeabilization is thought to be essential for promoting cell recovery (Gehl et al., 1999) and compounds such as Poloxamer 188 facilitate membrane resealing (Lee et al., 1992).

5. Conclusion

In conclusion, there are multiple potential ways to introduce trehalose into mammalian cells and in some cases excellent cell preservation can be achieved. However, it is clear that methods for each cell type will need to be diligently developed and many years of work remain before we can replace DMSO as the lead cryoprotectant. In the mean time, we must not forget that there are other relatively low molecular weight sugars available. Preliminary evidence suggests that with further work sucrose and stachyose may, in some cases, be equally effective for cell preservation.

6. Acknowledgements

We would like to thank Elizabeth Greene for her assistance in the preparation of this manuscript. This work was supported by a cooperative agreement (No. 70NANB1H3008) between the U.S. Department of Commerce, National Institute of Standards and Technology — Advanced Technology Program, and Organ Recovery Systems, Inc

7. References

Acker, J.P., Lu, X.M., Young, V., Cheley, S., Bayley, H., Fowler, A. & Toner M. (2003). Measurement of trehalose loading of mammalian cells porated with a metal-actuated switchable pore. *Biotechnology and Bioengineering*, Vol.82, No.5, (June 2003), pp. 525-32, PMID: 12652476.

Barnett, R. (1978). The effects of dimethyl sulfoxide and glycerol on Na+. K+-ATPase and membrane structure. *Cryobiology*, Vol.15, (April 1978), p. 227, PMID: 149651..

Bayley, H. (1994). Triggers and Switches in a Self-Assembling Pore-Forming Protein. *Journal of Cellular Biochemistry*, Vol.56, (October 1994), pp. 177-82, PMID: 7829577.

Beattie, G.M., Crowe, J.H., Lopez A.D., Cirulle, V., Ricordi, C. & Hayek, A. (1997). Trehalose: A cryoprotectant that enhances recovery and preserves function of human pancreatic islets after long-term storage, *Diabetes*, Vol.46, (March 1997), pp. 519-23, PMID: 9032112.

Brockbank, K.G.M., Campbell, L.H., Ratcliff, Kelly M. & Sarver, K.A. (2007, 2011). Method for treatment of cellular materials with sugars prior to preservation. U.S. Patents #7,270,946, #8,017,311.

Brockbank, K.G.M., Campbell, L.H., Greene, E.D., Brockbank, M.C.G. & Duman, J.G. (2011). Lessons Learned from Nature for Preservation of Mammalian Cells, Tissues and Organs, *In Vitro Cellular & Developmental Biology–Animal*, Vol.47, No.3, (March 2011), pp. 210-217, PMID: 21191664.

Buchanan, S.S., Gross, S.A., Acker, J.P., Toner, M., Carpenter, J.F., & Pyatt, D.W. (2004). Cryopreservation of Stem Cells Using Trehalose: Evaluation of the Method Using a Human Hematopoietic Cell Line, *Stem Cells and Development*, Vol.13, (June 2004), pp. 295–305, PMID: 15186725.

Buchanan, S.S.; Menze, M.A.; Hand, S.C.; Pyatt, D.W.& Carpenter, J.F. (2005). Cryopreservation of Human Hematopoietic Stem and Progenitor Cells Loaded with Trehalose: Transient Permeabilization via the ATP-Dependent P2Z Receptor Channel, *Cell Preservation Technology*, Vol. 3, No.4, (2005), pp. 212–22.

Buchanan, S.S., Pyatt, D.W. & Carpenter J.F. (2010). Preservation of Differentiation and Clonogenic Potential of Human Hematopoietic Stem and Progenitor Cells during Lyophilization and Ambient Storage, *PLoS ONE*, Vol.5, No.9, (September 2010), pp. e12518, PMID: 20824143.

Campbell, L.H., Taylor, M.J. & Brockbank, K.G.M. (2003) Two stage method for thawing cryopreserved cells, U.S. Patent #6,596,531.

Campbell, L.H., & Brockbank, K.G.M. (2007). Serum free solutions for the cryopreservation of cells, *In Vitro Cellular & Developmental Biology–Animal*, Vol.43, No.8-9, (September 2007), pp. 269-275, PMID: 17879124.

Campbell, L.H., Brockbank, K.G.M. (2011). Comparison of Electroporation and Chariot™ for Delivery of β-galactosidase into Mammalian Cells: Strategies to Use Trehalose in Cell Preservation, *In Vitro Cellular & Developmental Biology–Animal*, Vol.47, No.3, (March 2011), pp.195-199, PMID: 21184200.

Campbell, L.H., Sarver, K.A., Hylton, K.R., Leman, B.B., & Brockbank, K.G.M. (2011). Culturing with trehalose produces viable endothelial cells after cryopreservation, *Cryobiology*, Submitted, 2011.

Carpenter, J.F., Hand, S.C., Crowe, L.M., & Crowe, J.H. (1986). Cryoprotection of Phosphofructokinase with Organic Solutes: Characterization of Enhanced Protection in the Presence of Divalent Cations, *Archives of Biochemistry and Biophysics*, Vol.250, No.2, (November 1986), pp. 505-1, PMID: 2946263.

Carpenter, J.F., Crowe, L.M .& Crowe JH. (1987a). Stabilization of Phosphofructokinase with Sugars During Freeze-Drying: Characterization of Enhanced Protection in the Presence of Divalent Cations, *Biochimica et Biophysica Acta*, Vol.923, No.1, (January 1987), pp. 109-15, PMID: 2948571.

Carpenter, J.F., Martin, B., Crowe, L.M. & Crowe, J.H. (1987b). Stabilization of Phosphofructokinase During Air-Drying with Sugars and Sugar/Transition Metal Mixtures, *Cryobiology*, Vol.24, No.5, (October 1987), pp. 455-64, PMID: 2958239.

Carpenter, J.F. & Crowe, J.H. (1989). An Infrared Spectroscopic Study of the Interactions of Carbohydrates with Dried Proteins, *Biochemistry*, Vol.28, No.9, (May 1989), pp. 3916-22, PMID: 2526652.

Caselli, D., Tintori, V., Messeri, A., Frenos, S., Bambi, F. & Arico, M. (2009). Respiratory depression and somnolence in children receiving dimethyl sulfoxide and morphine during hematopoietic stem cell transplantation. *Haematologica*, Vol.94. No. 1, (January 2009), pp. 152–153, PMID: 19001279.

Crowe, L.M., Womersley, C., Crowe, J.H., Reid, D., Appel, L. & Rudolph, A. (1986). Prevention of Fusion and Leakage in Freeze-Dried Liposomes by Carbohydrates". *Biochimica et Biophysica Acta*, Vol.861, (1986), pp. 131-40.

Crowe, J.H., Crowe, L.M., Carpenter, J.F. & Aurell Wistrom, C. (1987). Stabilization of Dry Phospholipid Bilayers and Proteins by Sugars, *The Biochemical Journal*, Vol.242, No.1, (February 1987), pp. 1-10, PMID: 2954537.

Crowe, J.H., Crowe, L.M., Carpenter, J.F., Rudolph, A.S. & Aurell Wistrom, C., Spargo, B.J. & Anchordoquy, T.J. (1988). Interactions of Sugars with Membranes, *Biochimica et Biophysica Acta*, Vol.947, No.2, (June 1988), pp. 367-84, PMID: 3285894.

Crowe, J.H., Crowe, L.M. & Hoekstra, F.A. (1989a). Phase Transitions and Permeability Changes in Dry Membranes During Rehydration, *Journal of Bioenergetics & Biomembranes*, Vol.21, No.1, (February 1989), pp. 77-91.

Crowe, J.H., McKersie, B.B. & Crowe, L.M. (1989b). Effects of Free Fatty Acids and Transition. Temperature on the Stability of Dry Liposomes, *Biochimica et Biophysica Acta*, Vol.979, No.1, (February 1989), pp. 7-10, PMID: 2917168.

Crowe, J.H., Carpenter, J.F., Crowe, L.M. & Anchordoguy, T.J. (1990). Are Freezing and Dehydration Similar Stress Vectors? A Comparison of Modes of Interaction of Stabilizing Solutes with Biomolecules, *Cryobiology*, Vol.27, No.3, (June 1990), pp. 219-31.

Crowe, J.H. & Crowe, L.M. (1991). Preservation of Liposomes by Freeze Drying, In: *Liposome Technology*, G. Gregoriadis, CRC Press, ISBN 0849340764, Boca Raton, FL.

Crowe, J.H., Carpenter, J.F. & Crowe, L.M. "(1998). The role of Vitrification in Anhydrobiosis, *Annual Review of Physiology*, Vol.60, (1998), pp. 73-103, PMID: 9558455.

Crowe, J.H., Crowe, L.M., Oliver, A.E., Tsvetkova, N., Wolkers, W. & Tablin, F. (2001). The Trehalose Myth Revisited: Introduction to a Symposium on Stabilization of Cells in the Dry State, *Cryobiology*, Vol.43, No.2, (September 2001), pp. 89-105, PMID: 11846464.

Eroglu, A., Russo, M.J., Bieganski, R., Fowler, A., Cheley, S., Bayley, H. & Toner, M. (2000). Intracellular trehalose improves the survival of cryopreserved mammalian cells, *Nature Biotechnology*, Vol.18, No.2, (February 2001), pp. 163-67, PMID: 10657121.

Futaki, S., Suzuki, T., Ohashi, W., Yagami, T., Tanaka, S., Ueda. K., Sugiura, Y. (2001). Arginine-rich Peptides: An Abundant Source of Membrane-Permeable Peptides having Potential as Carriers for Intracellular Protein Deliver, *Journal of Biological Chemistry*, Vol. 276, No.8, (February 2000), pp. 5836-40, PMID: 11084031.

Futaki, S., Goto, S., Suzuki, T., Nakase, I. & Sugiura, Y. (2004). Structural Variety of Membrane Permeable Peptides, *Current Protein & Peptide Science*, Vol.4, No.2, (April 2004), pp. 87-96, PMID: 12678848.

Gehl, J., Sorensen, T.H. & Nielsen, K. (1999). In Vivo Electroporation of Skeletal Muscle: Threshold, Efficacy and Relation to Electric Field Distribution, *Biochimica et Biophysica Acta*, Vol.1428, No.2, (August 1999), pp. 233-40, PMID: 10434041

Gehl, L. (2003). Electroporation: theory and methods, perspectives for drug delivery, gene therapy and research, *Acta Physiologica Scandinavica*, Vol.177, No.4, (April 2003), pp. 437-47, PMID: 12648161

Hapala, I. (1997). Breaking the Barrier: Methods of reversible permeabilization of cellular membranes, *Critical Reviews in Biotechnology*, Vol.17, No.2, (1997), pp. 105-22, PMID: 9192473.

Heiser, W.C. (2000). Optimizing electroporation conditions for the transformation of mammalian cells, *Methods in Molecular Biology*, Vol.130, (2000), pp. 117-34, 2000. PMID: 10589426

Junior, A.M., Arrais, C.A., Saboya, R., Velasques, R.D., Junqueira, P.L. & Dulley, F.L. (2008). Neurotoxicity associated with dimethyl sulfoxide-preserved hematopoietic progenitor cell infusion. *Bone Marrow Transplantation*, Vol.41, No. 1, (January 2008), pp. 95-6, PMID: 17934528.

Katkov, I. (2002). Electroporation of cells in applications to cryobiology: summary of 20-year experience, *Problems of Cryobiology*, Vol.2, (2002), pp. 3-8.

Katkov, I., Isachenko, V., Isachenko, E., Kim, M., Lulat, A., Mackay, A. & Levine, F. (2006). Low- and high-temperature vitrification as a new approach to biostabilization of reproductive and progenitor cells, *International Journal of Refrigeration*, Vol.29, (February 2006), pp. 346-357.

Katkov, I., Kan, N., Cimadamore, F., Nelson, B., Snyder, E., Terskikh, A. (2011). DMSO-Free programmed cryopreservation of fully dissociated and adherent human induced pluripotent stem cells, *Stem Cells International*, Vol.2011, (June 2011), pp.1-8, PMID: 21716669.

Katsuda, S., Okada, Y., Nakanishi, I. & Tanaka, J. (1984). The influence of dimethyl sulfoxide on cell growth and ultrastructural features of cultured smooth muscle cells, *Journal Electron Microscopy*, Vol.33, No. 3, (July 1984), pp. 239-241, PMID: 6533232.

Katsuda, S., Okada, Y. & Nakanishi, I. (1987). Dimethyl sulfoxide induces microtubule formation in cultured arterial smooth muscle cells. *Cell Biology International Reports*, Vol.11, (February 1987), pp. 103-110, PMID: 3549003.

Langedijk, J.P.M. (2002), Translocation Activity of C-terminal Domain of Pestivirus E[rns] and Ribotoxin L3 Loop, *Journal of Biological Chemistry*, Vol.277, No.7, (February 2002), pp. 5308-14, PMID: 11673454.

Langedijk, J.P.M., Olijhoek, T., Schut, D., Autar, R. & Meloen, R.H. (2004). New Transport Peptides Broaden the Horizon of Applications for Peptidic Pharmaceuticals, *Molecular Diversity*, Vol. 8, No.2, (2004), pp. 101-11, PMID: 15209161

Lee, R.C., River, L.P., Pan, F.S. & Ji, L. (1992). Surfactant-Induced Sealing of Electropermeabilized Skeletal Muscle Membranes In Vivo, *Proceedings of the*

National Academy of Sciences USA, Vol.89, No.10, (May 1992), pp. 4524-28, PMID: 1584787

Lindgren, M., Hällbrink, M., Prochiantz, A., Langel, Ü. (2000). Cell-penetrating Peptides, *Trends in Pharmacological Sciences*, Vol.21, No.3, (March 2000), pp. 99-103, PMID: 10689363.

Lovelock, J. & Bishop, M., (1959). Prevention of freezing damage to living cells by dimethyl sulfoxide. *Nature*, Vol.183, (May 1959), p. 1394, ISSN: 0022-202X.

Lundberg, M. & Johansson, M. (2002). Positively Charged DNA-Binding Proteins Cause Apparent Cell Membrane Translocation, *Biochemical and Biophysical Research Communications*, Vol.291, No.1, (February 2002), pp. 367-71, PMID: 11846414.

Mai, J.C., Shen, H., Watkins, S.C., Cheng, T. & Robbins, P.D. (2002). Efficiency of Protein Transduction is Cell Type-dependent and is Enhanced by Dextran Sulfate, *Journal of Biological Chemistry*, Vol.277, No.33, (August 2002), pp. 30208-18, PMID: 12034749.

Mathew, A.J., Baust, J.M., Van Buskirk, R.G. & Baust, J.G. (2004). Cell preservation in reparative and regenerative medicine: evolution of individualized solution composition, *Tissue Engineering*, Vol.10, No. 11, (November 2004), pp. 1662-71, PMID: 15684675.

Miranda, A.F., Nette, G., Khan, S., Brockbank, K.G.M. & Schonberg, M. (1978). Alteration of myoblast phenotype by dimethyl sulfoxide, *Proceedings of the National Academy of Sciences USA*, Vol.75, (August 1978), pp. 3826-3830, PMID: 278996.

Mondal, B. (2009). A simple method for cryopreservation of MDBK cells using trehalose and storage at −80°C, *Cell and Tissue Banking*, Vol.10, No.4, (November 2009), pp. 341-344, PMID: 19381873.

Morris, M.C., Robert-Hebmann, V., Chaloin, L., Mery, J., Heitz, F., Devaux, C., Goody, R.S. & Divita G. (1999). A New Potent HIV-1 reverse transcriptase Inhibitor. A Synthetic Peptide Derived from the Interface Subunit Domain, *Journal of Biological Chemistry*, Vol.274, No.35, (August 1999), pp. 24941-6, PMID: 10455170.

Morris, M.C., Depollier, J., Mery, J., Heitz, F., Divita, G. (2001). A Peptide Carrier for the Delivery of Biologically Active Proteins into Mammalian Cells, *Nature Biotechnology*, Vol.19, No. 12, (December 2001), pp. 1173-76, PMID: 11731788.

Mueller, L.P., Theurich, S., Christopeit, M., Grothe, W., Muetherig, A., Weber T, Guenther, S. & Behre, G. (2007). Neurotoxicity upon infusion of dimethyl sulfoxide-cryopreserved peripheral blood stem cells in patients with and without pre-existing cerebral disease. *European Journal of Haematology*, Vol.78, No.6, (June 2007), pp. 527-31, PMID: 17509106.

Murgia, M., Pizzo, P., Steinberg, T.H. & Di Virgilio, F. (1992). Characterization of the Cytotoxic Effect of Extracellular ATP in J774 Mouse Macrophages, *The Biochemical Journal*, Vol.288, (December 1992), pp. 897-901, PMID: 1472003.

Mussauer, H., Sukhorukov, V.L. & Zimmermann, U. (2001). Trehalose Improves Survival of Electrotransfected Mammalian Cells, *Cytometry*, Vol.45. No.3, (November 2001), pp. 161-9, PMID: 11746084

Nihei, O.K., Savino, W. & Alves, L.A. (2000). Procedures to Characterize and Study $P2_Z/P2_{X7}$ Purinoceptor: Flow Cytometry as a Promising Practical, Reliable Tool,

Memorias do Instituto Oswaldo Cruz, Vol.95, No.3, (May 2000), pp. 415-28, 2000, PMID: 10800201.

Oliver, A.E., Jamil, K., Crowe, J.H. & Tablin, F. (2004) Loading Human Mesenchymal Stem Cells with Trehalose by Fluid-Phase Endocytosis, *Cell Preservation Technology*, Vol.2, No.1, (2004) pp. 35-49.

Otrock, Z.K., Beydoun, A., Barada, W.M., Masroujeh, R., Hourani, R. & Bazarbachi A. (2008). Transient global amnesia associated with the infusion of DMSO-cryopreserved autologous blood stem cells. *Haematologica*, Vol.93, No.3, (March 2008), pp. 36–7, PMID: 18310533.

Polge, C., Smith, A.U. & Parkes A. (1949). Revival of spermatozoa after vitrification and dehydration at low temperatures. *Nature*. Vol.164, (October 1949), p. 666, ISSN: 0028-0836

Richard, J.P., Melikov, K., Vives, E., Ramos, C.,Birgit, V., Gait, M.J., Chernomordik, L.V. & Lebleu, B. (2003). Cell-penetrating Peptides: A Reevaluation of the Mechanism of Cellular Uptake, *Journal of Biological Chemistry*, Vol. 278, No.1, (January 2003), pp. 585-90, PMID: 12411431.

Rudolph, A.S. & Crowe, J.H. (1985). Membrane Stabilization During Freezing: The Role of Two Natural Cryoprotectants, Trehalose and Proline, *Cryobiology*, Vol.22, No.4, (August 1985), pp. 367-77, PMID: 4028782.

Schlegel, P.G., Wölfl, M., Schick, J., Winkler, B. & Eyrich, M. (2009). Transient loss of consciousness in pediatric recipients of dimethyl sulfoxide (DMSO)-cryopreserved peripheral blood stem cells independent of morphine co-medication, *Haematologica*, Vol.94, No.10, (October 2009), pp. 1473-5, PMID: 19608681.

Slade, L. & Levine, H. (1991). A food polymer science approach to structure-property relationships in aqueous food systems: non-equilibrium behavior of carbohydrate-water systems, *Advances in Experimental Medicine and Biology*, Vol.302, (1991), pp. 29-101, PMID: 17416335.

Sosef, M.N., Baust, J.M., Sugimachi, K., Fowler, A., Tompkins, R.G. & Toner, M. (2005) Cryopreservation of isolated primary rate hepatocytes: enhanced survival and long-term hepatospecific function, *Annals of Surgery*,.Vol.241, No.1, (January 2005), pp. 125-33, PMID: 15622000.

Taylor, M., Campbell, L., Rutledge, R., Brockbank, K. (2001). Comparison of Unisol with Euro-collins solution as a vehicle solution for cryoprotectants, *Transplantation Proceedings*, Vol.33, (February 2001), pp. 667-9, PMID : 11267013.

Vives, E., Brodin, P. & Lebleu, B. (1997). A Truncated HIV-1 Tat Protein Basic Domain Rapidly Translocates through the Plasma Membrane and Accumulates in the Cell Nucleus, *Journal of Biological Chemistry*, Vol.272, No.25, (June 1997), pp. 16010-17, PMID: 9188504.

Walker, B., Braha, O., Cheley, S. & Bayley, H.(1995). An intermediate in the assembly of a pore-forming protein trapped with a genetically-engineered switch. *Chemistry & Biology*, Vol.2, No,2, (February 1995), pp. 99-105, PMID: 9383410.

Zeng, X., Zhao, C., Wang, H., Li, S., Deng, Y. & Li, Z. (2010). Dimethyl Sulfoxide Decrease Type-I and –III Collagen Synthesis in Human Hepatic Stellate Cells and Human

Foreskin Fibroblasts. *Advanced Science Letters*, Vol.3, No. 4, (December 2010), pp. 496–499.

4

Cryopreservation – A Viable Alternative in Preparation for Use of Allografts in Knee Ligament Reconstruction

Alexandre C. Bitar et al.*
Vita Institute
Brazil

1. Introduction

The use of allogenic tissues is growing in orthopedic practice, as well as the number of studies on methods for processing, sterilization and cryopreservation that interfere as little as possible with the original physiological properties of the tissues (Nutton et al., 1999).

In addition to bone tissue, other tissues of the locomotor system can be captured, processed and stored in tissue banks with the purpose of transplantation. Therefore a strict quality control must be implemented and set after discussions compiled by international organizations such as AATB e EATB.

The first report of the use of allografts in humans dates back to 1881. The first tissue bank of bone grafts was created in 1940 in the United States and the initial clinical results were published in 1942 by Inclan, 1942. Since then a series of regulations and studies has emerged relating to the use of grafts in orthopedic practice[19]. Currently, tendon allografts are used in knee surgeries, in elbow ligament reconstructions and in revisions of the acromioclavicular joint (Costic et al., 2004).

In our country we have few tissue banks. The tissue bank (BTME) from the Institute of Orthopedics and Traumatology (IOT), Hospital das Clínicas da Faculdade de Medicina da Universidade de Sao Paulo was the first and is the biggest bank in activity nowadays. It thas been in operation since 1999 and is governed by local legislation (Amatuzzi et al., 2000). With the restructuring of our service in 2005, we initiated a new technique aimed at the provision of the tendon with a well-structured quality program in line with other centers of excellence nationally and internationally. Today our service is provided by a series of ' tendons (tibial tendon, Achilles, patellar and peroneal) taken from different regions with very specific applications. Thus, we can then follow the technological trend in the use of

*Caio Oliveira D'Elia[1], Antônio Guilherme P. Garofo[1], Wagner Castropil[1], Luiz Augusto U. Santos[2], Marco K. Demange[2], José Ricardo Pécora[2] and Alberto T. Crocci[2]
[1]Vita Institute, Brazil
[2]Universidade de São Paulo; Institute of Orthopedics and Traumatology, Hospital das Clinicas da Faculdade de Medicina da Universidade de São Paulo, SP, Brazil

tissues already practiced in other global centers of excellence in orthopedics and traumatology.

In our medical service, Vita Institute (private use), and in the Institute of Orthopedics and Traumatology (public and private use), allografts are used mainly in knee surgeries, ACL reconstruction, multiple ligament reconstructions, ligament surgery in skeletally immature patients and with double bundle reconstruction.

2. Importance of allografts in knee ligament reconstructions

Grafts are used in various procedures in different branches of orthopedics, including ligament reconstruction. The literature shows the importance of the use of allografts in knee surgery, especially in revision surgeries, multiple ligament reconstructions and, more recently, surgery for primary reconstruction of the anterior cruciate ligament (ACL) in active patients and in those aged over 40 years (Marrale et al., 2007; Sherman & Banffy, 2004). There have been at least 11 clinical studies comparing the use of auto and allografts in the reconstruction of the ACL (Chang et al., 2003; Marrale et al., 2007). Most of these show little difference between the two techniques with respect to long-term results. However, there has been few prospective randomized studies , and the comparison methods (scores), types of graft, as well as methods of preparing and fixing the graft are highly variable.

Furthermore, most studies use the patellar tendon graft; therefore, it may not be possible to generalize the conclusions of these studies to the flexor tendon. Some authors, such as Lawhorn and Howell, suggest the use of allografts without a bone plug because of the potential for slower incorporation of the bone due to immunogenicity and smaller cross-sectional area of transplants with bone plugs (Lawhorn & Howell, 2003). Recently, Sun et al. (2011) published a prospective randomized study comparing non-irradiated allograft with flexors autograft showing similar results between groups in terms of subjective clinical scores, goals, rate of return to sports and incidence of complications (Sun et al., 2011).

However, compared with autologous transplants, allografts do have some advantages. For example, they do not increase morbidity for the donor, they require a shorter surgery time, and they are available without restriction on size and morphology. In ligament reconstruction surgeries, the possibility exists of an immune response from the recipient tunnel enlargement, and delayed incorporation of the allograft (Marrale et al., 2007). The risk of disease transmission and the potential for immunogenicity are the major disadvantages of allografts, but these complications can be controlled (Albert et al., 2006; Barrios et al., 1994, Urabe et al., 2007).

3. Controversy of methods of preparation of allografts

The increased frequency of the use of allografts in traumato-orthopedics requires the adoption of storage techniques that interfere as little as possible in the quality of the parts (Vangsness et al., 2003). Allografts can be stored in different ways; they can be chilled in residential mechanical freezers at temperatures of + 2°C to - 4°C for up to five days. In freezers with temperatures of -20°C to -40°C, they can be stored for up to six months[14]. At these temperatures, the enzymes present in the tissue are still active and can destroy the tissue. Therefore, storage periods of longer than a few months are not recommended. The methods of sterilization used at low temperatures are effective against fungi and do not seem to change the mechanical characteristics of the grafts. The period of 40 days chosen in our service for cryopreservation coincides with the period of incubation necessary for microbiological investigations for bacteria and fungi (Vangsness et al., 2003). The deep-freezing process enables storage for up to five years, and this is the method we use in our service[19].

Many services prefer to carry out the manipulation of tissues under aseptic conditions from acquisition through clinical use, and the samples are discarded when microbiological assays show positive bacterial cultures (20 to 30%) (Zimmerman et al., 1994). Sterilization methods, therefore, are not completely safe. They can alter the biomechanical characteristics of tissues or fail to penetrate tissue layers, resulting in the protection of microorganisms rather than their destruction. Irradiation with gamma rays is the most common method of sterilization (Sterling et al., 1995). However, to achieve safer sterilization in frozen tissues, high-dose irradiation is necessary, which can alter the biomechanical properties of the tissue in a dose-dependent manner (Curran et al., 2004; Fideler et al., 1995).

Doses as low as 2 Mrad resulted in a statistically significant reduction in biomechanical properties, outcomes, or physical examination measures. Rappe et al. (2007) studied the effect of irradiation on clinical outcomes of ACL reconstruction, they found the irradiated group had an unacceptable higher rate of failure (33%) than the non- irradiated group (2.4%). Fideler et al. (1995) found that the dose of 2.5 Mrad, which was a dose commonly used by tissue banks for sterilization, was just bacteriocidal but ineffective in eliminating viruses such as human immunodeficiency virus (HIV) (Sterling et al., 1995). Doses of 3 to 4 Mrad were necessary to inactivate the virus. Grieb et al. also proved that lower levels of radiation may be inadequate to kill hepatitis and HIV viruses, with a dose of 5 Mrad being necessary (Grieb et al., 2006). When dosage is increased, its clinical implications increase correspondingly. We must question the use of gamma irradiation as there are so many adverse effects and it fails to sterilize the allograft as required.

Also, the sterilization effectiveness against viruses is low (Vangsness et al. 2003). Ethylene oxide sterilization requires strict control of the levels of waste gas in contact with the allograft and is no longer used by tissue banks, due to the possibility of toxic effects for the recipient (dissolution of the graft and articular inflammatory reactions) (Vangsness et al. 2003). The processing techniques used in the preparation and preservation of grafts have been questioned as potentially altering the initial resistance and mechanical properties of the graft prior to implantation.

Two studies carried out in Brazil address the biomechanical properties of patellar and calcaneus tendons of cadavers with the same preparation method as that used in our study, comparing fresh and cryopreserved allografts (Giovani et al., 2006; Reiff et al., 2007). They found no differences. A study on metric measurements and attachment levels of the medial patellofemoral ligament shows this to be a distinct structure (Zimmerman et al., 1994). Although there have been studies on the biomechanical behavior of tendons, the literature does not address histological changes of tendons cryopreserved at -80°C under aseptic conditions (Pearsall et al., 2003). During cryopreservation at -80°C, thedestruction of the allograft enzyme appears to be minimal and at least one enzyme, collagenase, which can destroy the tissue, is inactive (Tomford, 1997). Furthermore, with cryopreservation there is no intracellular free water, which is thought to be necessary for enzymatic activity, bacterial proliferation and lipid oxidation (Galea & Keamey, 2005; Laitinien et al., 2006). Lipid oxidation inside the tissues induces apoptosis and inhibits cell differentiation; such oxidation can be minimized or avoided with cryopreservation at temperatures of at least -70° C (Laitinien et al., 2006). The literature refers to histological changes due to cryopreservation only in cartilage (one of the most commonly used grafts in surgical practice), concluding that during freezing, the vitality of the cells is threatened (Schchar & McGann, 1986). Other injuries may also occur, such as the formation of extracellular ice crystals, intracellular ice nucleation, collapse of the matrix, and breakage of intercellular bridges.

In our study, the histological study of one tendon (not cartilage) was carried out, and none of these histological phenomena were observed with cryopreservation at -80°C. Freezing with liquid nitrogen at -179°C has also been used as a storage method with similar results but higher cost (Zimmerman et al., 1994). Another widespread storage method is lyophilization. Cryopreservation and lyophilization have been related to a reduction in allograft antigenicity (Jackson et al., 1990). The use of chilled saline solution is not a guaranteed method because the stock can only be kept safely for short periods (Zimmerman et al., 1994). Treatment with paraformaldehyde and fixation with glutaraldehyde are no longer recommended because of the toxicity of these solutions to the recipient tissue.

We recently published a study in which we proved the histological properties of the flexor tendons of the knee from cadavers subjected to cryopreservation and experience with the use of allografts of the Knee Group from IOT (Bitar et al., 2010; Damasceno et al., 2009).

4. Cryopreservation: Our method

4.1 Tendinous tissue removal

The attainment of the musculoskeletal tissues has as its source deceased donors with brain death reported by the Committees Intra-hospital - CIHDOTs, Organ Procurement Organizations - OPOS and by the twenty-three central of notification and collection of organs and tissues - CNCDOs, logistically spread throughout country. Notifications for teams pickups are made after the execution of a series of procedures and tests that aim beyond the evidence of brain death, family consent of the donation of organs and tissues.

The donor selection follows a rigorous research with control antigen and antibody serology for HIV, Hepatitis A, B and C, HTLV-1 and 2, syphilis, Chagas disease, toxoplasmosis and cytomegalovirus in addition to testing of last generation for evidencing of DNA (Nucleic Acid Amplification - NAT) for HIV and Hepatitis B and C, required of musculoskeletal tissues. The capture of musculoskeletal tissues (bone and tendons) is performed after the initial screening of donors of multiple organs and tissues (heart, kidney, liver, pancreas, lung, cornea, etc.).

In our specific field, we follow a protocol of evaluation of the donor that counts with a written anamnesis of a term to capture and physical examination. Are excluded donors with orthopedic disorders such as osteoporosis, osteonecrosis, rheumatoid arthritis, lupus erythematosus, malignancy, age that compromises characteristic of tissues, blood transfusion, tattoos or adornments (piercings) within the window period, users of illicit drugs, permanence in endemic areas, generalized or localized infections, fractures, bruises on the limbs which are absorbed in the musculoskeletal tissues or any other situation that would call into doubt the quality of these tissues, as arranged in the existing laws. The tissues removed are immediately packed in triple enclosures, hermetically sealed and sent under refrigeration (- 4 ° C) to the Tissue Bank.

A very important step of the process of capture is the reconstruction of the donor and for this matter we use prosthesis specially designed using plaster, wire suture, gauze. This reconstruction is done rigorously and is characterized as the most laborious phase of the procedure. All anatomical parameters are respected, and therefore the deformation of the donor does not occur (Figures 1 and 2).

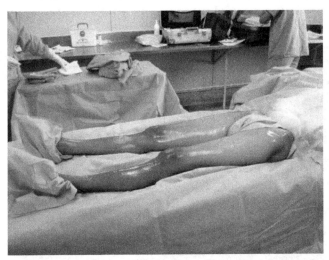

Fig. 1. Pre-operative preparation of the potential musculoskeletal tissues doner.

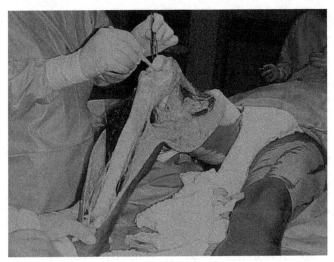

Fig. 2. Tissue removal: bone and tendon dissecation under asseptic conditions.

4.2 Processing and cryopreservation of musculoskeletal tissues

At the end of the uptake, the tissues are sent to BTME chilled in coolers with temperature monitoring throughout the period of transportation. The processing step is preceded by a planning of activities needed for its implementation, such as provision of materials and instruments, convocation of the processing team, definition of preparation and dimensioning according to the need for service (queue) requests from orthopedic and dental surgeons. This step is performed in the operating room properly rated (class 100 or ISO 5) equipped with integrated laminar flow (Figure 3). The room also has an antechamber and pass-through and all environments have strict control of air particles and positive pressure for quality assurance of tissues processed there.

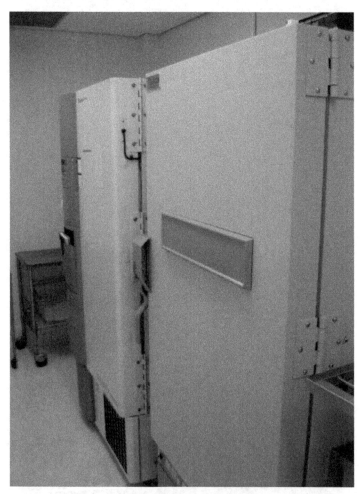

Fig. 3. Tissue cryopreservation área. Ultrafreezer with a temperature of -80°C.

In addition, specific attire is required of the professional team that should only use non-woven clothes to avoid dispersion of particles that emit the cotton clothes (Figure 4). Not only the non-woven attire is required but the team's behavior should be differentiated. Thus, sudden movements, use of cosmetics and hair exposure should be avoided during the permanence in this room. Ensuring an appropriate approach is not only a result of training of the nursing staff that performs the procedure, but also other professionals who access the environment (cleaning maintenance).

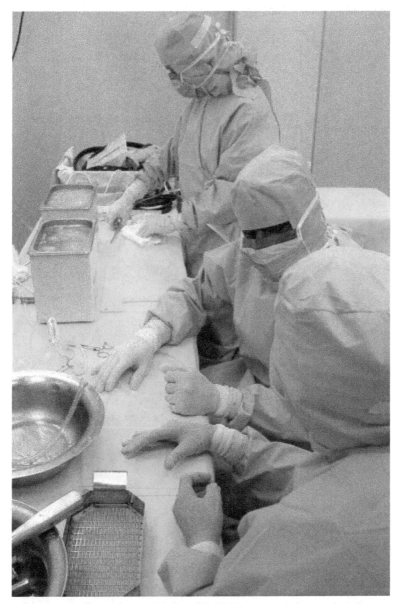

Fig. 4. Processing team in activity in the controlled área (ISO 5: Class 100).

The BTME conducts various types of processing of these tissues with the purpose to use in orthopedic and dental surgeries, each of which requires a specific plan. For the processing of fresh frozen tissues it is performed what we call mechanical processing, ie removal of adventitial tissue such as blood, periosteum, subcutaneous tissue, muscle, fascia and fibrotic tissue. Then, these tissues are immersed in emulsifying solutions based on hydrogen peroxide and alcohol under ultrasonic agitation (Figure 5).

Fig. 5. Tissue's chemical processing. Ultrasonic clining with emulsifying solution.

Then, a sampling of these resultant solutions, of bone marrow of long bones and fragments of each tissue processing are subjected to microbiological examination (General Knowledge, and Culture of Anaerobic Fungi). Furthermore, it is also obtained samples to histopathological analysis.

Finally starts the procedure of packaging of all the processed grafts which are measured (length, height, weight, volume, perimeter) and kept in sterile triple wrappers, vacuum sealed and properly identified as tissue in analysis. The label contains information from the donor, examination, lot number, item, expiration date, type of conservation and barcode.

Once all the tissues are identified, they are x-rayed at the very BTME and referred to cryopreservation.

The bones can also be processed in its lyophilized form, where all water is removed with the tissue still frozen. The process involves placing the tissue in a lyophilizer chamber where ice crystals sublimate by the action of the high pressure, not passing through the liquid phase and thus maintaining the viability of bone matrix. The result is a dry tissue, conservable at room temperature that must receive final sterilization by irradiation.

At the end of the processing it is performed the documentation of the procedure in the Processing and archived in the donor's chart. The stock of tissues can be kept either frozen or dried, if necessary, according to the same standards used by the Global Association of Tissue Banks . Other forms of processing have been investigated in order to reduce costs related to storage and maintenance. The glicerolization of bone tissue is presented as a processing methodology capable of maintaining the viability of the matrix and prevent bacterial growth, and allows storage at room temperature (Giovani et al., 2006).

4.3 Tissue cryopreservation

In the room of cryopreservation tissues are stored according to their status in the process. Thus, there is a space for tissues in analysis or in quarantine (where they remain for about 60 days until the results of all examinations) and those already released for use. Both rooms are equipped with ultrafreezers with temperatures ranging from 85 to 110 degrees below zero (Amatuzzi et al., 2000).

The room is also equipped with air-conditioning system, own power generator and the unfreezing protection of carbon dioxide (CO2 Backup), and a rigorous system of monitoring the temperature, with printed record of temperature for 24 hours and alarm system via satellite, which guarantees the right temperature and early detection of complications.

Depending on the outcome of the analysis, the tissues are transferred to the room of material released for use. The maximum period of cryopreservation is five years to bone tissue and two years for soft tissues and tendons.

4.4 Quality control and distribution

By the time of transplantation, all tissues processed are subjected to rigorous quality assurance criteria. It requires the evaluation of all data pertinent to the donor, test results, maintenance and control of equipment, materials and instruments used in all phases of each procedure (Figures 6 and 7).

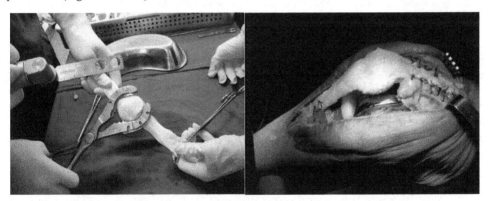

Fig. 6. and 7. Patellar tendon allograft transplant.

All processes are computerized through the System Manager of the Tissue Bank, a program designed to record all the steps which allows the rescue and traceability of each graft processed and delivered. Through a coding is possible to identify the donor, lot, expiration, and status of the tissue examination (under review, released, deleted and used).

Given the need of retrieval of information, as the evidence of an adverse effect, you can quickly and safely obtain the information and implementation of corrective and preventive actions.

For a lot of graft in analysis to be released for use nurses must analyze the results of all tests performed: NAT serology or PCR for HIV, HBC and HCV, General Culture, Anaerobic Culture and Culture of Fungi, pathology reports and radiographic findings. These reports of examinations are assessed and ultimately released by the Technical Director of the Tissue Bank.

Besides examinations, evaluation of the printed record of temperature during the storage period is considered. The temperature oscillations are quickly detected and reported to the team members from the BTME even remotely by cell phones. In addition, audible alarms at strategic points in the hospital and the presence of CO_2 backups, ensure system reliability.

After the release of each lot, the nurses carry a detailed examination of integrity of each tissue during the replacement of tissues in analysis labels to released labels and posterior transfer of the sector. The logistics of storage of tissues in ultrafreezers considers the type of tissue and speed up the search.

We emphasize that for a rigorous quality control all steps of each procedure are carried out through check-lists with double checking and approval.All relevant data of the donor or lot records are filed in a single file and stored at the BTME for a minimum of 25 years.

A serum bank with plasma samples of donors are offered by BTME if necessary examination of counterproof.

5. Our experience

In the last five years, we have performed 35 knee ligament reconstructions, including multiple ligament, and isolated ACL and PCL reconstructions with single and double bundle techniques, and ACL reviews. Twenty seven men and eight women underwent surgery with the employment of the following grafts from the tissue bank: anterior tibial tendon (48 units), patellar tendon (4 units), quadriceptal tendon (5 units), semitendinosus tendon (4 units), calcaneus tendon (1 units) and fibular tendon (1 units). The patients follow-up range from 6 to 57 months and we are still collecting data from this cases. Our first results showed that there no viral or bacterial infection associated to the use of allografts in any of the cases or other complications, and clinical outcome of these patients has shown good results with the use of tendons from the tissue bank.

A study conducted in another service, in which we colaborate, revised the records of 46 patients who were submitted to ligament reconstructions between 1999 and 2007 using grafts supplied by the same tissue bank (Damasceno et al., 2009). Thirty-four male patients and 12 female patients were reviewed, with follow-up time ranging from 10 months to 9 years (mean: 3.1 years).The surgical procedures used 9 units of patellar tendons, 9 units of anterior tibial tendons, 8 units of calcaneal tendons, 6 units quadriceptal tendons and 1 unit of fibular tendon, mainly for multiple ligamentar reconstructions and ACL reviews[41]. There were also no viral or bacterial infection cases[41].

The decrease in morbidity and postoperative complications allied to good results obtained in our samples reinforces the idea that the use of allografts is a good and safe option in knee ligament reconstruction.

6. References

Albert, A; Leemrijse, T; Druez, V; Delloye, C; Cornu, O. (2006) Are bone autografts still necessary in 2006? A three-year retrospective study of bone grafting. *Acta Orthop Belg*, 72, 734-40. ISSN 0001-6462.

Amatuzzi, MM; Croci, AT; Giovani, AMM; Santos, LAU. (2000) Banco de tecidos: estruturação e normatização. [Tissue bank: structure and organization]. *Rev Bras Ortop*, 35, 165-72. ISSN 0102-3616.

Barrios, RH; Leyes, M; Amillo, S; Oteiza, C. (1994) Bacterial contamination of allografts. *Acta Orthop Belg*, 60, 293-5.

Bitar, AC; Santos, LAU; Croci, AT; Pereira, JARM; França Bisneto, EM; Giovani, AMM; Oliveira, CRGCM. (2010) Histological Study of Fresh versus Frozen Semitendinous Muscle Tendon Allografts. *CLINICS*, 65(3),297-303 ISSN 1807-5932.

Chang, SK; Egami, DK; Shaieb, MD; Kan, DM; Richardson, AB. (2003) Anterior cruciate ligament reconstruction: allograft versus autograft. *Arthroscopy*, 19, 453-62. ISSN 0749-8063.

Costic, RS; Labriola, JE; Rodosky, MW; Debski, RE. (2004) Biomechanical rationale for development of anatomical reconstructions of coracoclavicular ligaments after complete acromioclavicular joint dislocations. *Am J Sports Med*, 32, 1929-36. ISSN 1552-3365.

Curran, AR; Adams, DJ; Gill, JL; Steiner, ME; Scheller, AD. (2004) The biomechanical effects of low-dose irradiation on bone-patellar tendon-bone allografts. *Am J Sports Med*, 32, 1131-1135. ISSN 1552-3365.

Damasceno, ML; Ferreira, TF; D'Elia, CO; Demange, MK; Pécora, JR; Hernandez, AJ; Camanho, GL; Croci, AT; Santos, L; Helito, CP. (2009) Use of Allograft in Ligamentar Reconstruction of Knee. *Acta Ortop Bras*, 17(5), 265-8 ISSN 2176-7521.

Fideler, BM; Vangsness, CT Jr; Lu, B; Orlando, C; Moore T. (1995) Gamma irradiation: effects on biomechanical properties of human bone-patellar tendon-bone allografts. *Am J Sports Med*, 23, 643-646. ISSN 1552-3365.

Flahiff, CM; Brooks, AT; Hollis, JM; Vander Schilden, JL; Nicholas, RW. (1995) Biomechanical analysis of patellar tendon allografts as a function of donor age. *Am J Sports Med*, 23, 354-8. ISSN 1552-3365.

Galea, G; Kearney, JN. (2005) Clinical effectiveness of processed and unprocessed bone. *Transfus Med*. 15:165-74.

Giovani, AM; Croci, AT; Oliveira, CR; Filippi, RZ; Santos, LA; Maragni, GG. (2006) Comparative study of cryopreserved bone tissue and tissue preserved in a 98% glycerol solution. *CLINICS*, 61, 565-70. ISSN 1807-5932.

Grieb, TA; Forng, RY; Bogdansky, S. (2006) High-dose gamma irradiation for soft tissue allografts: high margin of safety with biomechanical integrity. *J Orthop Res*, 24, 1011-1018. ISSN 0736-0266.

Inclan A. (1942) The use of preserved bone graft in orthopaedic surgery. *J Bone Joint Surg Am*, 24, 81-96.

Jackson, DW; Corsetti, J; Simon TM. (1996) Biologic incorporation of allograft anterior cruciate ligament replacements. *Clin Orthop*, 324, 126-33.

Jackson, DW; Windler, GE; Simon, TM. (1990) Intraarticular reaction associated with the use of freeze-dried, ethylene oxide-sterilized bone-patella tendon-bone allografts in the reconstruction of the anterior cruciate ligament. *Am J Sports Med*, 18, 1-10. ISSN 1552-3365.

Laitinen, M; Kivikari, R; Hirn M. (2006) Lipid oxidation may reduce the quality of a fresh-frozen bone allograft. Is the approved storage temperature too high? *Acta Orthop*, 77, 418-21. ISSN 1745-3674.

Lawhorn, KW; Howell, SM. (2003) Scientific justification and technique for anterior cruciate ligament reconstruction using autogenous and allogenic soft-tissue grafts. *Orthop Clin North Am*, 34, 19-30. ISSN 0030-5898.

Levitt, RL; Malinin, T; Posada, A; Michalow, A. (1994) Reconstruction of anterior cruciate ligaments with bone-patellar tendon-bone and achilles tendon allografts. *Clin Orthop Relat Res*, 303, 67-78. ISSN 0009-921X.

Marrale, J; Morrissey, MC; Haddad, FS. (2007)A literature review of autograft and allograft anterior cruciate ligament reconstruction. *Knee Surg Sports Traumatol Arthrosc*, 15, 690-704. ISSN 0942-2056.

Nutton, RW; McLean, I; Melville, E. (1999) Tendon allografts in knee ligament surgery. *J R Coll Surg Edinb*, 44, 236-40.

Papandrea, P; Vulpiani , MC; Ferretti, A; Conteduca, F. (2000) Regeneration of the semitendinosus tendon harvested for anterior cruciate ligament reconstruction. Evaluation using ultrasonography. *Am J Sports Med*, 28, 556-61. ISSN 1552-3365.

Pearsall, AW; Hollis, JM; Russell, GV Jr; Scheer, Z. (2003) A biomechanical comparison of three lower extremity tendons for ligamentous reconstruction about the knee. *Arthroscopy*, 19, 1091-6. ISSN 0749-8063.

Rappe, M; Horodyski, M; Meister, K; Indelicato, PA. (2007) Nonirradiated versus irradiated Achilles allograft: in vivo failure comparison. *Am J Sports Med*, 35, 1653-1658. ISSN 1552-3365.

Reiff, RBM; Croci, AT; Bolliger Neto, R; Pereira, CAM. (2007) Estudo comparativo de propriedades biomecânicas da porção central do tendão calcâneo congelado e a fresco. [Comparative study on biomechanical properties of the central portion of frozen and fresh calcaneus tendon]. *Acta Ortop Bras*, 15, 6-8. ISSN 2176-7521.

Schachar, NS; McGann, LE. (1986) Investigations of low-temperature storage of articular cartilage for transplantation. *Clin Orthop Relat Res*, 146- 50. ISSN 0009-921X.

Sherman, OH; Banffy, MB. (2004) Anterior cruciate ligament reconstruction: which graft is best? *Arthroscopy*, 20, 974-80. ISSN 0749-8063.

Sterling, JC; Meyers, MC; Calvo RD. (1995) Allograft failure in cruciate ligament reconstruction. Follow-up evaluation of eighteen patients. *Am J Sports Med*, 23, 173-8. ISSN 1552-3365.

Sun, K; Zhang, J; Wang, Y; Xia, C; Zhang, C; Yu, T; Tian, S. (2011) Arthroscopic Reconstruction of the Anterior Cruciate Ligament With Hamstring Tendon Autograft and Fresh-Frozen Allograft A Prospective, Randomized Controlled Study. *Am J Sports Med*. [Epub ahead of print] ISSN 1552-3365.

Tomford, W. Transmission of disease through musculoskeletal transplantation. *Portland Bone Symposium*. Portland: Oregon Health Sciences University, 1997.

Urabe, K; Itoman, M; Toyama, Y; Yanase, Y; Iwamoto, Y; Ohgushi, H. (2007) Current trends in bone grafting and the issue of banked bone allografts based on the fourth nationwide survey of bone grafting status from 2000 to 2004. *J Orthop Sci*, 12, 520-525. ISSN 1436-2023.

Vangsness, CT Jr; Garcia, IA; Mills, CR; Kainer, MA; Roberts, MR; Moore, TM. (2003)Allograft transplantation in the knee: tissue regulation, procurement, processing, and sterilization. *Am J Sports Med*, 31,474-481. ISSN 1552-3365.

Zimmerman, MC; Contiliano, JH; Parsons, JR; Prewett, A; Billotti, J. (1994) The biomechanics and histopatology of chemically processed patellar tendon allografts for anterior cruciate ligament replacement. *Am J Sports Med*, 22, 378-386. ISSN 1552-3365.

5

Validation of Primary Packaging for Cryopreserved Musculoskeletal Tissues

Luiz Augusto Ubirajara Santos[1], Rosa Maria Vercelino Alves[2], Alberto T. Croci[1], Fábio Gomes Teixeira[2], Paulo Henrique Kiyataka[2], Marisa Padula[2], Mary Ângela Fávaro Perez[2], Monica Beatriz Mathor[3], Renata Miranda Parca[6], Arlete M.M. Giovani[1], Cesar Augusto Martins Pereira[4], Graziela Guidoni Maragni[1], Thais Queiróz Santolin[1] and Lucas da Silva Pereira[5]

1. Introduction

Bones and tendons are obtained from donors who have been pronounced brain dead after a rigid and extensive screening process. The tissues obtained are sent to a Tissue Bank and submitted to processing steps, packaging and cryopreservation at -80°C. It is vital to maintain sterility and integrity, so that no tissue is discarded.These precautions also extend to the packaging, which should promote containment and protection. Although minimum technical criteria have already been defined for the food industry, this has still not been regularized in Brazil. We emphasize, therefore, the need to study this subject, focusing on maintaining the quality of the musculoskeletal tissues produced by tissue banks. All procedures developed byTissue Banks currently present in the country have rigid control over the quality and the traceability of tissues made available, based on international standards (EATB, 2004; AATB, 2007) on the legislation (BRASIL, 1997-2006) and in conformity with Good Manufacturing Practices - GMP.

2. Cryopreservation of musculoskeletal tissues

In the cryopreservation room, the tissues are stored according to their status in the process. Thus, there are designated areas for tissues under analysis or in quarantine (where they remain for around 60 days before the result of all the exams) and for those that have been liberated for use. both areas are equipped with ultra-low temperature freezers, with temperatures ranging from minus 80 to minus 100 degrees Celsius.

[1]Institute of Orthopedics and Traumatology, Hospital das Clínicas of the University of São Paulo School of Medicine, Sao Paulo, SP, Brazil
[2]Packaging Technology Center – Institute of Food Technology, Campinas, SP, Brazil
[3]Nuclear and Energy Research Institute - IPEN/CNEN-SP – São Paulo, Brazil
[4]Biomechanics Laboratory - Institute of Orthopedics and Traumatology, Hospital das Clínicas of the University of São Paulo school of Medicine, Brazil
[5]Dental Student Institute of Orthopedics and Traumatology, Hospital das Clínicas of the University of São Paulo school of Medicine, São Paulo, SP, Brazil
[6]National Health Surveillance Agency- ANVISA

The room is also equipped with an air conditioning system, its own energy generator, and carbon gas backup as protection against defrosting, as well as a rigorous temperature monitoring system that generates 24-hour printed temperature recordings, and a satellite alarm system, which ensures adequate maintenance of the temperature and early detection of any flaws.

Depending on the results of the analyses, the tissues are transferred to the room designated for materials liberated for use. The maximum cryopreservation period is 5 years for bone tissue and 2 years for soft or tendinous tissues.

It is vital to maintain sterility and integrity, in order to avoid disposal of any material. This is the purpose of **packaging**, which is aimed at containment and protection. In Brazil, there are no specific regulations for packaging of sterile, cryopreserved tissues for transplants, and in this chapter, we present our experience in the definition and validation of a type of packaging used for this purpose. Our proposal is to characterize a coextruded plastic film structure used for packaging of musculoskeletal tissues at low temperatures, in regard to the aspects sterility, cytotoxicity, migration, mechanical resistance and oxygen permeability.

Fig. 1. Cryopreservation of tissues at -80 °C in drawers, separated by batches.

3. Validation of packaging for use in tissue cryopreservation.

3.1 Description of our sample

A roll of transparent, unprinted seven layers, coextruded plastic:

Layer 01: LLDPE – 1-octene comonomer Linear Low-Density Polyethylene
Layer 02: LLDPE – 1-octene comonomer Linear Low-Density Polyethylene and 1-octene comonomer Linear Low-Density Polyethylene modified with anhydride maleic
Layer 03: PA – hexamethylenediamine, adipic acid e caprolactam copolyamide
Layer 04: LLDPE – 1-octene comonomer Linear Low-Density Polyethylene and 1-octene comonomer Linear Low-Density Polyethylene modified with anhydride maleic
Layer 05: PA – hexamethylenediamine, adipic acid e caprolactamcopolyamide
Layer 06: LLDPE – 1-octene comonomer Linear Low-Density Polyethylene and 1-octene comonomer Linear Low-Density Polyethylene modified with anhydride maleic
Layer 07: LLDPE – 1-octene comonomer Linear Low-Density Polyethylene

In relation to the physicomechanical and barrier properties characteristics, we investigated the alterations that occurred in the packages after 30, 60, 90, 120, and 150 days of cryopreservation at -80°C. For the cytotoxicity and sterility analysis, two groups (before and after sterilization by ethylene oxide) were analysed.

3.2 Physicomechanical and barrier properties tests

A) Characterization of coex film in relation to its thickness and water vapor transmission rate (WVTR)

The thickness of each layer of plastic material in the film sample was determined through images captured by a Metaval inverted microscope operating at a magnification of 200x, using the image analysis system Axio Vision (Zeiss®). The cross-section of the sample was obtained using a Leica microtome, model RM2245, with section thickness set to 40 μm. To facilitate visualization of the barrier layer, 2% iodine solution was used as a contrasting agent. Five cross-section samples of the material were obtained; five measurements were performed for each sample, totalling 25 thickness measurements. The test was carried out at a temperature of around 23°C, after packaging of the sample in a controlled environment at 23°C ± 2°C and (50 ± 3)% relative humidity for a minimum period of 48 hours.

The water vapor transmission rate was determined using a MOCON PERMATRAN-W 3/31 device, following the procedure described in regulation ASTM F1249-06 - Standard test methods for water vapor transmission rate through plastic film and sheeting using a modulated infrared sensor. In this test, the water vapor that passes through the film is carried to the infrared sensor by ultra-dry nitrogen flow. The sensor measures the fraction of energy absorbed by the water vapor, and emits an electrical signal with amplitude proportional to the concentration of water vapor. The range of this signal is compared with that of the signal produced by the water vapor that passes through a calibration film with a known water vapour transmission rate. The effective permeation area of each sample was 50cm². The assay was performed at 38°C/100%RH and in this condition, the calibration standard showed water vapour transmission rate of 4.54g water.m^{-2}.day^{-1}. The water vapor transmission rate of the sample was corrected for the condition 38°C/90%RH, multiplying the results by a factor of 0.9.

The total thickness of each layer of coextruded film is shown in Table 1. Figure 10 shows an example of a cross-section image of the sample obtained for the assay.

Determinations	Thickness (μm)		
	Mean	Variation Interval	Coefficient of variation (%)
Total	91.1	90.0 – 92.9	0.9
LLDPE (A)	27.3	26.2 – 28.5	2.4
LLDPE blend (B)	7.6	7.0 – 8.0	3.8
PA (C)	5.6	5.1 – 5.9	4.8
LLDPE blend (D)	6.8	5.3 – 7.8	11
PA (C)	5.6	5.0 – 6.0	5.6
LLDPE blend (F)	7.7	7.1 – 8.6	5.9
LLDPE (G)	31.0	30.1 – 32.1	1.9

Values for 25 measurements
A-G: Text for visualization in Figure 10

Table 1. Total thickness of each layer of sample of coextruded plastic film.

Thus, by optical microscopy, it was observed that it is a coextruded material with seven layers, with two intermediate layers of PA of approximately 6μm each. The other LLDPE layers totalled 80μm.

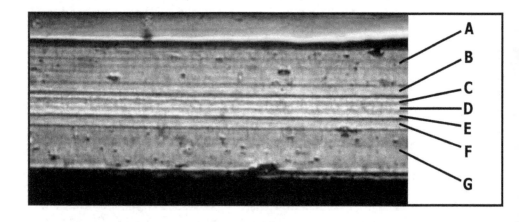

Fig. 2. Example of cross-section image of the sample obtained with the microscope operating at a magnification of 200x.

The characterization of the film in relation to water vapor transmission rate is shown in Table 2.

Sample	WVTR (g water . m-2.day-1)		
	Mean	VI	CV (%)
Coextruded film LLDPE / LLDPE / PA / LLDPE / PA / LLDPE / LLDPE	3.30	3.2 – 3.4	4.4

Values relate to four measurements
VI – variation interval; CV – coefficient of variation

Table 2. Water vapor transmission rate (WVTR) of the coextruded plastic film at 38°C/90%RH – Permatran-W 3/31 method.

The WVTR depends on the thickness of the LLDPE layers of the coex film (80μm).

B) Penetration resistance

The penetration resistance of the coextruded film was determined based on Standard ASTM F 1306-90 (2008) e1 - Standard test method for slow rate penetration resistance of flexible barrier films and laminates, on an Instron 5500R universal testing machine, using load cells of 100 N. The speed of penetration, performed with a spherical-tipped metal probe with diameter of approximately 3.2 mm, was 25 mm/min. The penetration was performed from the inner surface to the outer surface of the material. The test was conducted in an environment of 23°C ± 2°C and (50 ± 3)% relative humidity, after leaving the packaged samples for a minimum period of 48 hours in this environment.

The evaluation of penetration resistance was carried out on samples in their original or non-frozen condition (0) and 30, 60, 120 and 150 days after freezing at -80 °C. The data are shown in Table 3 and Figure 3.

Penetration resistance		Evaluation periods (days)					
		0	30	60	90	120	150
Force at break (N)	Mean	12.1 [ab]	12.6 [bc]	12.9 [c]	12.6 [bc]	11.1 [d]	11.8 [a]
	VI	11.5-12.7	11.8– 13.5	11.9– 14.2	11.2– 14.1	9.4 – 12.3	10.9– 12.4
	CV (%)	3.1	3.6	4.7	6.9	9.1	3.7
Deformation at break (mm)	Mean	9.4 [a]	11.0 [b]	10.0 [c]	9.8 [c]	8.3 [d]	9.0 [e]
	VI	9.2 – 9.8	10.7– 11.4	9.5 – 10.5	8.9 – 10.6	7.4 – 9.1	8.2 – 9.5
	CV (%)	2.3	2.3	2.6	5.2	5.5	3.7
Energy at break (mJ)	Mean	70 [ad]	76 [b]	78 [b]	73 [ab]	62 [c]	69 [d]
	VI	64 – 74	70 – 84	72 – 85	64 – 86	52 – 73	62 – 73
	CV (%)	4.6	5.5	5.4	8.6	11	5.6

Values relate to 10 measurements: 1 N = 0.102 kgf
VI/CV: Variation interval/Coefficient of variation
a,b,c,d: for an analyzed property, mean values on the same line accompanied by the same superscript letter did not show any difference between them in the least significant difference (LSD) test, at a level of error of 5%.

Table 3. Penetration resistance of the sample during storage at -80°C.

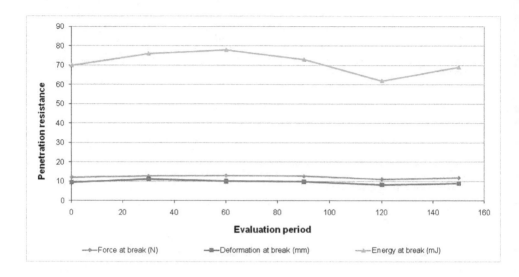

Fig. 3. Penetration resistance of the sample during the period of storage at -80°C (days).

The results shown in Table 3 and Figure 3 demonstrate that the process of sterilization and packaging of human bone, and storage for 150 days at -80°C, did not alter the penetration properties of the coextruded film. Variations were observed in the results of the three properties evaluated, including statistical differences in some periods/properties, but as these variations were small, without any clearly-defined trend, no alteration is expected in penetration resistance of the film during packaging and storage of human tissue for a period of 150 days at -80 °C.

C) Seal strength

The seal strength of the packaging was determined according to ASTM F 88/F 88M-09 - Standard test method for seal strength of flexible barrier materials. Samples of 25.4 mm in width were submitted to tensile test in a 5500R Instron universal testing machine, operating with load cells of 50N and 100N, at a speed of 300mm/min. The distance between the fixing clamps and the sample was 25mm. The test was conducted in an environment of 23 °C ± 2 °C and (50 ± 3)% relative humidity, after leaving the pre-prepared, packaged samples in this same environment for at least 48 hours.

The strength of the side seal is different from that of the top seal, because the top seal of the packaging was made after the bone tissue packaging process, while the side seals were made by the package manufacturer.

Despite this, there was no statistical difference between the results obtained for both seals. Thus, the sterilization and packaging process of human tissue, and storage for 150 days at -80°C did not alter the seal strength.

Maximum seal strength		Evaluation periods (days)					
(kgf/25.4 mm)		0	30	60	90	120	150
Side seal	Mean	3.60 ab	3.65 b	3.60 ab	3.73 b	3.73 b	3.44 a
	VI	3.36 – 3.95	3.41 – 3.83	3.38 – 3.80	3.34 – 3.97	3.43 – 3.96	3.12 – 3.78
	CV (%)	4.4	4.0	4.4	5.7	4.5	7.0
Top seal	Mean	-	4.69 a	4.75 a	5.07 a	5.09 a	4.84 a
	VI	-	3.65 – 5.20	3.20 – 5.85	4.31 – 5.95	4.68 – 5.54	4.48 – 5.44
	CV (%)	-	9.6	15	9.6	5.1	6.6

Values relate to ten determinations: 1 kgf/25.4 mm = 386.1 N/m
VI/CV: Variation interval/Coefficient of variation
a,b: for a seal type, mean values on the same line accompanied by the same superscript letter did not show any difference between them in the least significant difference (LSD) test, at a level of error of 5%.

Table 4. Seal strength during storage at -80°C.

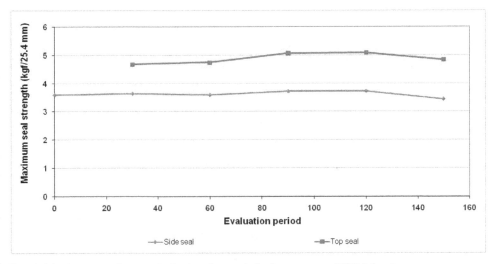

Fig. 4. Maximum seal strength during the period of storage at -80°C (days).

D) Oxygen transmission rate

The oxygen transmission rate at humid was determined by the coulometric method, according to the procedure described in ASTM F 1927 - Standard test method for determination of oxygen gas transmission rate, permeability and permeance at controlled relative humidity through barrier materials using a coulometric detector, on MOCON OXTRAN device, model 2/20, operating with pure oxygen as permanent gas. The tests were carried out at 23°C to 75%RH, with the samples packaged for 88 to 112 hours in a temperature-controlled room at 25°C and 75%RH. The effective area of permeation of each sample was 50cm². The result obtained was corrected for 1 atm of partial pressure gradient of oxygen.

O_2TR mL (STP).m^{-2}.day^{-1} at 23°C/75%RH	Evaluation periods (days)					
	0	30	60	90	120	150
Mean	69.75 [a]	75.95 [a]	74.97 [a]	82.57 [a]	76.98 [a]	113.67 [b]
VI	68.63– 70.87	71.77- 80.14	74.48- 75.46	79.34- 85.80	73.83- 80.13	105.05- 122.29
CV (%)	2.3	7.8	0.9	5.5	5.8	10.7

values relate to two determinations
VI – variation interval; CV – coefficient of variation
a,b: for a seal type, mean values on the same line accompanied by the same superscript letter did not
show any difference between them in the least significant difference (LSD) test, at a level of error of 5%.

Table 5. Oxygen transmission rates (O_2TR) during storage at -80°C.

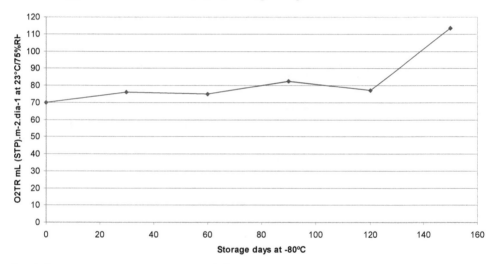

Fig. 5. Oxygen transmission rates (O_2TR) during storage at -80°C.

The results of the oxygen permeability rates shown in Table 5 and Figure 5 indicate a mean increase of 10% in permeability after the process of sterilization, vacuum packaging of the bone, and storage at -80°C, which was observed after 30 days of storage. This tendency to increase, probably due to humidification of the PA, led to a small loss of barrier which was maintained throughout the storage period of 120 days. Meanwhile, in the analysis of samples from 30 days of storage at -80°C, a high oxygen transmission rate of was observed, which was not expected, and as we did not have any more stored samples, it was not possible to re-evaluate this result. This higher O2TR may be the result of some variation in thickness of the PA in the samples evaluated in this period. In any case, the level of oxygen transmission rate of the film did not lead to loss of vacuum in the samples stored in the Tissue Bank for [1]50 days at -80°C.

E) Overall Migration

The evaluations of overall migration were performed according to Resolution 51 of 26 November 2010 published by *Agência Nacional de Vigilância Sanitária do Ministério da Saúde* (National Agency of Sanitary Surveillance of Health Ministry) in Diário Oficial da União (official journal of the Brazilian Government) on 30 November 2010. This Resolution internalizes Mercosul Technical Regulation GMC 32/10.

The methodology to quantify the overall migration was according to the method of the European Standard EN 1186-1: materials and articles in contact with foodstuffs. Plastics. Part 1: guide to the selection of conditions and test methods for overall migration, EN 1186-3: Materials and articles in contact with foodstuffs. Plastics. Test methods for overall migration into aqueous food simulants by total immersion and EN 1186-14: materials and articles in contact with foodstuffs. Plastics. Part 1: Part 14: test methods for "substitute tests" for overall migration from plastics intended to come into contact with fatty foodstuffs using test media isooctane and 95% ethanol and consists of the sample contact with extraction solutions in certain periods and temperatures that simulate their actual condition of use. The residues of overall migration were determined by the weight difference after the contact and evaporation of the solutions through an analytical scale with 0.01mg of accuracy. The sample was evaluated under the contact conditions shown in Table 6.

Model Solution	Contact condition
Ultra purified water	40ºC/10 days
Acetic acid solution in Ultra purified water at 3% (w/v)	40ºC/10 days
Isooctane	40ºC/10 days

Table 6. Conditions of time and temperature used in the overall migration.

The results of the overall migration tests performed on the transparent coextruded plastic film, obtained using the model solutions and the specific contact conditions are shown in Table 7.

Model Solution/Contact Condition	Maximum limit of overall migration	Sample	Mean	Standard Deviation	Variation interval
Ultra purified water/ 40ºC/10 days	8.0	Before EtO	≤ 0.72	0.21	≤ 0.50[2] - 1.00
		After EtO	≤ 0.51	0.02	≤ 0.50[2] - 0.55
3% Acetic acid solution (w/v)/ 40ºC/10 days	8.0	Before EtO	≤ 0.54	0.09	≤ 0.50[2] – 0.67
		After EtO	0.92	0.19	0.72 – 1.12
Isooctane/ 20ºC/48 hours	8.0	Before EtO	≤ 1.21	0.64	≤ 0.50[2] – 1.85
		After EtO	≤ 1.50	0.72	≤ 0.50[2] – 2.05

(1) Result of four determinations.
(2) Limit of quantification of the method in the analytical conditions used.
(3) Not applicable.

Table 7. Residues of overall migration obtained for the transparent coextruded plastic film, before and after the application of EtO, in mg/dm2 [1].

The maximum limit of overall migration provided by Resolution n°105/99 is of 8mg of residue per dm^2 of contact plastic material, with an analytical tolerance of 10%. Therefore 8.8mg/dm^2 is the maximum tolerable value.

The overall migration values found in the samples analyzed, in the analytical conditions used, were below the established limit. There was no statistical significance with sterilization with ethylene oxide (EtO).

E 1.) Physicochemical tests – According to USP 33

The physicochemical tests were conducted based in the methodology describe in the Chapter <661> Containers – Plastics - Physicochemical Tests of the **United States Pharmacopeia (USP 33)**.

In accordance to the United States Pharmacopeia, physicochemical tests are designed to determine physical and chemical properties of plastic materials. The extracts methodology consists of the sample contact with a extraction solution (deionized water) at 70°C during 24 hours, maintaining the ratio area / volume of 120 cm^2 total surface area of plastic material for each 20 mL of extraction solution.

The analyzed sample was received in the form of the film cut into strips with dimensions of 5.0 cm long, 0.3 cm wide and thickness less than 0.1 cm. In this case, the thickness of the material to determine the total area was not considered and was maintained the ratio of 120 cm^2 for each 20 mL of extraction solution. Water was used as extraction solution.

After the contact, the extraction solution and blank reagent were analyzed by the following tests:

Buffering Capacity: Titrate 20 mL of the extraction solution potentiometrically to a pH of 7.0, using 0.01 N sodium hydroxide. Treat a 20.0 mL portion of the blank reagent similarly. The difference between the two volumes can not be greater than 10.0 mL.

Nonvolatile Residue: 50 mL of the extraction solution were evaporated on a hot plate, after the residue was dried at 105°C for 1 hour on a oven and finally the nonvolatile residue was weighted through an analytical balance with 0.01mg of accuracy. Treat a 50.0 mL portion of the blank reagent similarly. The difference between the two volumes can not be greater than 15.0 mg.

Residue on Ignition: in the residues obtained in nonvolatile residues test, add sulfuric acid and burn on the muffle furnace until constant weight. Treat the blank reagent similarly. The difference between the two volumes should not be greater than 5.0 mg. It is not necessary to perform this test when the nonvolatile residue test result does not exceed 5.0 mg.

Heavy Metal, as lead: an aliquot of extraction solution has been transferred to a volumetric flask and acidified with nitric acid and the volume was completed with the extraction solution. After treating the sample, the lead content was quantified by atomic emission spectrometry induced by plasma, with an optical detector, in a Perkin Elmer equipment, model OPTIMA 2000DV, using appropriate calibration curves for the analyses. This test was conducted in replacement to the heavy metal test stablished by American Pharmacopeia, whose result is expressed as lead (and is based on the colour comparison among the test solution and a solution of lead with concentration of 1.0 mg / kg).

Physicochemical Assay	Limit based on USP	Sample	Mean	Standard deviation	Variation interval
Buffering capacity (mL)	10.0 mL	Before EtO	≤ 0.5[2]	[3]	[3]
		After EtO	≤ 0.5[2]	[3]	[3]
Non-volatile residue: (mg)	15 mg	Before EtO	≤ 1.0[2]	[3]	[3]
		After EtO	≤ 1.0[2]	[3]	[3]
Heavy Metals (as lead) (mg/kg (ppm))	1 mg/kg (ppm)	Before EtO	≤ 0.05[2]	[3]	[3]
		After EtO	≤ 0.05[2]	[3]	[3]

(1) Results of three determinations.
(2) Corresponds to the limit of quantification of the method in the analytical conditions used.
(3) Values no applicable

Table 8. Physicochemical assays of the analyzed samples[1].

E 2.) Assays According to the European Pharmacopoeia

The tests acidity or alkalinity, sulfated ash, absorbance, extractable aluminum, chromium, titanium, vanadium, zinc, zirconium and extractables from heavy metals, expressed as lead were conducted based in the methodology described in the European Pharmacopoeia, Chapters "3.1.3 Polyolefines", "3.1.4 Polyethylene without Additives for Containers for Parenteral Preparations and for Ophthalmic Preparations" and "3.1.5 Polyethylene with Additives for Containers for Parenteral Preparations and for Ophthalmic Preparations".

Extractables of aluminium, chromium, titanium, vanadium, zinc and zirconium

The methodology for quantification of extractables aluminum, chromium, titanium, vanadium, zinc and zirconium involved contact of 100 grams of sample with a solution of 0.1 M hydrochloric acid for one hour at the reflux temperature. After treating the sample, the metals contents were quantified by atomic emission spectrometry induced by plasma, with an optical detector, in a Perkin Elmer equipment, model OPTIMA 2000DV, using appropriate calibration curves for the analyses.

Extractables of Heavy Metals, expressed as lead

The same procedure described for quantification of extractable aluminum, chromium, titanium, vanadium, zinc and zirconium was used. The method for quantification of extractable lead was used instead of the colorimetric method of determination of heavy metals established by the European Pharmacopoeia, which is based on color comparison between the extracting solution and a solution of lead at a concentration of 2.5 mg/kg.

Alkalinity or Acidity, and Absorbance

The method for quantification of alkalinity or acidity and absorbance involved contact of 12.5 grams of sample with 250 mL of deionized water for five hours at the reflux temperature.

After the extraction time, both the extracting solution in contact with the samples and a blank solution (reference) were assessed as to the following tests:

Alkalinity and Acidity: measurement of pH in a Micronal pHmeter, model B 474 and titration of 100 mL of extracting solution with sodium hydroxide 0.01M or hydrochloric acid 0.01M up to pH 7.0.

Absorbance: absorbance of the extracting solution was measured in the 220 nm to 340 nm wavelength range, using a quartz cuvette with 10 mm pathlenght in a UV / VIS spectrophotometer, using an Analytik Jena instrument, model Specord 210.

Sulphated Ash

Sulfated ash were assessed per requirements in European Pharmacopoeia 6.0, Chapter 2.4.14 - Sulfated Ash. The method for quantification of sulfated ash consisted in weighing 5.00 g ± 0.01 g of sample on an analytical scale with 10^{-5} g resolution and incineration at a temperature of 600 °C ± 20 °C using a Milestone microwave heating furnace, model Pyro. After incinerating the sample, the ash were determined gravimetrically, using an analytical scale with 10^{-5} g resolution. There was not need to use sulfuric acid.

Extractable aluminum, chromium, titanium, vanadium, zinc and zirconium

The results of the extractable aluminum, chromium, titanium, vanadium, zinc and zirconium tests for the analyzed sample are shown in Table 9.

Extractables	Limit based on European Pharmacopeia 6.0	Sample	Mean	Standard Deviation	Variation interval
Al	1.0	Before EtO	0.71	0.08	0.62 – 0.78
		After EtO	0.74	0.06	0.70 – 0.81
Cr	0.05	Before EtO	0.03	0.00	0.02 – 0.03
		After EtO	0.33	0.05	0.27 – 0.37
Ti	1.0	Before EtO	≤ 0.10[2]	[3]	[3]
		After EtO	≤ 0.10[2]	[3]	[3]
V	0.10	Before EtO	≤ 0.10[2]	[3]	[3]
		After EtO	≤ 0.10[2]	[3]	[3]
Zn	1.0	Before EtO	0.05	0.00	0.05 – 0.06
		After EtO	0.05	0.01	0.05 – 0.06
Zr	0.10	Before EtO	≤ 0.05[2]	[3]	[3]
		After EtO	≤ 0.05[2]	[3]	[3]

(1) Result of four determinations.
(2) Corresponds to the limit of quantification of the equipment in the analytical conditions used.
(3) Not applicable.

Table 9. Extractables of the metals aluminum (Al), chromium (Cr), titanium (Ti), vanadium (V), zinc (Zn) and zirconium (Zr), in mg/kg[1].

Extractable Heavy Metal, Expressed as Lead

The results of the extractable heavy metals, expressed as lead, acidity, absorbance and sulphated ash tests, for the analyzed sample are shown in Table 10.

Physicochemical Assay	Limit based on European Pharmacopeia 6.0	Sample	Mean	Standard Deviation	Variation interval
Extractables of Heavy Metals, expressed as lead (ppm)	2.5mg/kg (ppm)	Before EtO	≤ 0.1[2]	[3]	[3]
		After EtO	≤ 0.1[2]	[3]	[3]
Acidity (mL of NaOH 0.01 M)	1.5 mL de sodium hydroxide 0.01 M	Before EtO	≤ 0.5[2]	[3]	[3]
		After EtO	≤ 0.5[2]	[3]	[3]
Absorbance (UA)	0.2 UA	Before EtO	0.23	0.01	0.22 - 0.23
		After EtO	0.24	0.04	0.22 - 0.26
Sulphated Ash (%)	0.02% and 1.00%[4]	Before EtO	0.29	0.01	0.28 - 0.30
		After EtO	0.25	0.12	0.11 - 0.33

UA- Unit of Absorbance.
(1) Results of three determinations.
(2) Corresponds to the limit of quantification of the method in the analytical conditions used.
(3) Values no applicable
(4) Varies according to the presence of additives

Table 10. Results for heavy metals, expressed as lead, acidity, absorbance and ash of the samples analyzed [1].

The lead values, volume of 0.1 M sodium hydroxide used, and sulphated ash values found in the samples analyzed were below the maximum limits established in the European Pharmacopeia 6.0. In relation to absorbance, the values obtained in the two samples analyzed, before and after the application of sterilization with ethylene oxide (EtO), were slightly higher than the maximum limits established.

A038-2/11 – Packaging for human tissue – Final Report 17/20 This means that some substance of the coextruded film may have migrated to the extraction solution in contact with the sample (deionized water), a fact that requires further investigation.

It should be emphasized that the methodology of the American and European Pharmacopeias apply to single-layer packaging, and that substances from internal layers of the film analyzed may have been extracted, slightly increasing the absorbance of the extraction solution.

E 3.) Specific migration

The extruded film before and after sterilization with ethylene oxide was evaluated in relation to specific migration of 1-octene and ε-caprolactam. The evaluations of specific migrations of 1-octene, ε-caprolactam and hexamethylenediamine were carried out according to the Brazilian legislation.

E 3.1) Specific migration of 1-octene

The quantification of specific migration of 1-octene was evaluated based on Standard CEN/ TS 13130-26: materials and articles in contact with foodstuffs – Plastics substances subject to limitation – Part 26:

Determination of 1-octene and tetrahydrofuran in food simulants, and consists of contact of the sample with solutions of extraction with times and temperatures that simulate its real condition of use.

The internal sides of the samples were placed in contact with the simulants, obeying an area:volume ratio of 600 cm2 to 1000 mL. The same was evaluated in the contact conditions shown in Table 11.

Simulants/Contact Condition	Maximum limit of specific migration of 1-octene	Sample	Mean [1]	Standard Deviation	Variation Interval
Ultra purified water/ 40°C/10 days	15	Before EtO	≤2.8[2]	[3]	[3]
		After EtO	≤2.8[2]	[3]	[3]
3% Acetic acid solution in ultra purified water (w/v)/ 40°C/10 days	15	Before EtO	≤1.6[2]	[3]	[3]
		After EtO	≤1.6[2]	[3]	[3]
Olive oil/ 40°C/10 days	15	Before EtO	≤8.7[2]	[3]	[3]
		After EtO	≤8.7[2]	[3]	[3]

(1) Result of three determinations.
(2) Quantification Limit of the method under the analytical conditions.
(3) Not applicable.

Table 11. Specific migration of 1-octene obtained for transparent coextruded plastic film, before and after the application of EtO, in mg/dm2.

The specific limit of monomer migration of 1-octene established in Resolution 105/99 of the National

Health Surveillance Agency – ANVISA of 19 May 1999 is 15 mg/kg of simulant. The values for specific migration of 1-octene found in the samples analyzed, in the analytical conditions used, were below the established limit. Sterilization with ethylene oxide (EtO) did not affect the monomer migration potential of 1-octene.

E 3.2) Specific migration of ε-caprolactam

The quantification of specific ε-caprolactam was evaluated based on Standard **CEN/ TS 13130-16**: materials and articles in contact with foodstuffs – Plastics substances subject to limitation - Part 16: Determination of caprolactam and caprolactam salt in food simulants.

The internal surfaces of the samples were placed in contact with the simulants, obeying an area:volume ratio of 600 cm2/1000 mL. The samples were evaluated under the contact conditions shown in Table 12.

The specific limit of monomer migration of ε-caprolactam established in Resolution 105/99 of the National Health Surveillance Agency – ANVISA of 19 May 1999 is 15 mg/kg of simulant. The values for specific migration of ε-caprolactam found in the samples analyzed, in the analytical conditions used, were below the established limit. Sterilization with ethylene oxide (EtO) did not affect the potential monomer migration of ε-caprolactam for the fatty simulant, but was significantly lower (probability of 95% confidence – Tukey Test) for the ultrapure water simulants and 3% acetic acid solution (m/v) in ultrapure water after sterilization with ethylene oxide.

Simulants/Contact Condition	Maximum limit of specific migration of ε-caprolactam	Sample	Mean [1]	Standard Deviation	Variation Interval
Ultra purified water/ 40°C/10 days	15	Before EtO	2.4	0.3	1.9 - 2.7
		After EtO	1.7	0.2	1.3 - 1.9
3% Acetic acid solution in ultra purified water (w/v)/ 40°C/10 days	15	Before EtO	2.7	0.1	2.6 - 2.8
		After EtO	2.2	0.3	1.8 - 2.6
Olive oil/ 40°C/10 days	15	Before EtO	3.4	0.4	3.0 - 4.0
		After EtO	3.2	0.2	2.9 - 3.5

(1) Result of three determinations.

Table 12. Specific migration of ε-caprolactam obtained for transparent coextruded plastic film, before and after the application of EtO, in mg/dm².

E 3.3) Specific migration of hexamethylenediamine

The quantification of specific hexamethylenediamine migration was evaluated based on Standard CEN/ TS 13130-21: materials and articles in contact with foodstuffs - Plastics substances subject to limitation – Part 21: Determination of ethylenediamine and hexamethylenediamine in food simulants. The sample of hexamethylenediamine, adipic acid e caprolactam copolyamide in the form of a film, was placed in contact with the simulants, obeying an area:volume ratio of 600 cm2 to 1000 mL. The sample was evaluated under the contact conditions shown in Table 13.

Simulants/Contact Condition	Maximum limit of specific migration of hexamethylenediamine	Mean [1]	Standard Deviation	Variation Interval
Ultra purified water/ 100 °C/30 minutes + 40 °C/10 days	2,4	≤1.2[2]	[3]	[3]
3% Acetic acid solution in ultra purified water (w/v)/ 100 °C/30 minutes + 40 °C/10 days	2,4	≤1.1[2]	[3]	[3]
Olive oil/ 100 °C/30 minutes + 40 °C/10 days	2,4	≤2.0[2]	[3]	[3]

(1) Result of three determinations.
(2) Quantification Limit of the method under the analytical conditions.
(3) Not applicable.

Table 13. Specific migration of hexamethylenediamine obtained for a 42 μm film copolyamide, in mg/kg.

The specific limit of monomer migration of hexamethylenediamine established in Resolution 105/99 of the National Health Surveillance Agency – ANVISA of 19 May 1999 is 2.4 mg/kg of simulant. The values for specific migration of hexamethylenediamine found in the analysis of the 42 μm film copolyamide, under the analytical conditions used, were below the established limit.

3.3 Cytotoxicity assay

Assay performed in accordance with Standard ISO 10993-5: Biological evaluation of medical devices – Part 5: Cytotoxicity Assays: *in vitro* methods in samples of coextruded plastic film.

3.3.1 Cytotoxicity

Definitions of cytotoxicity vary, depending on the nature of the study and whether cells are killed or simply have their metabolism altered. Cytotoxicity is the toxicological effect that a substance can cause *in vitro*, at cellular level (Freshney, 2000).

As defined in ISO 10993, "the numerous methods used and end-points measured in cytotoxicity determination can be grouped into categories of evaluation type, like assessments of cell damage by morphological means; measurements of cell damage; measurements of cell growth and measurements of specific aspects of cellular metabolism" [IOS, 2010]. The cells can be exposed to the samples or their extracts.

3.3.2 The protocols

Cell culture

Chinese hamster ovary cell line (CHO-k1) was standardized for cytotoxicity and genotoxicity tests. Cells were maintained in RPMI medium supplemented with antibiotics and antimycotics (100 units/mL penicillin, 100 μg/mL streptomycin and 0.025 μg/mL amphotericin), 2mM glutamine, and 10% calf serum, at 37º C in a humidified 5% CO_2 atmosphere until they reached confluence. For subculturing and for experiments, cells were harvested using 0.05% trypsin and 0.02% EDTA in phosphate-buffered saline, pH 7.4.

Extract preparation

Samples of packaging, before and after sterilization, were submitted to this assay. The samples were immersed separately in RPMI culture medium at a final concentration of 1 cm^2 / mL and left in an incubator at 37ºC for 72 hours to fulfil the extraction condition. The first concentration was sterilized by filtration and the subsequent dilutions were performed in sterile RPMI medium at a ratio of 1:2.

Cytotoxicity test

A colorimetric method that uses the tetrazolium compound MTS was used to determine the number of viable cells in proliferation (Cory et al, 1991). 96-well microplates were prepared with 50 μL of extract diluted from 100 to 6.25% in RPMI medium in quadruplicates. The positive control was a phenol solution (0.5% v/v) as 100% concentration and the negative control was a high density polyethylene (HDPE) extract. The 100% concentration was the non-extract well A suspension of CHO-k1 (from second to fourth passages after thawing) with 6 x 10^4 cell/mL was prepared and 50 µL/well was pipetted into the microplates. The microplates were incubated for 72 hours at 37ºC in a humidified 5% CO_2 atmosphere. Blank and controls of the cells were also prepared. Cell viability was measured by adding 20 µL of MTS/PMS (20:1) solution to the humidified 5% CO_2 incubator, followed by incubation for 2 hours at 37ºC. The microplates were read in a spectrophotometer reader at 490 nm.

Cell viability was calculated by the equation:

$$CV\% = \frac{OD\,sample}{OD\,non\,extract} \times 100$$

Where: CV% = cell viability, OD sample = optical density at 490 nm of the extract dilution, OD non extract = optical density at 490 nm of the well without extract.

The results consider the following parameters:

a.   Controls (positive and negative)

Positive control: 0.5% Phenol solution

Negative control: HDPE (high-density polyethylene) extract.

b.   Observations: definitions of some terms.
- Positive control: material which, when tested according to Standard ISO 10993-5, promotes a cytotoxic response.
- Negative control: material which, when tested according to Standard ISO 10993-5, does not promote a cytotoxic response.
- $IC_{50(\%)}$: cytotoxicity index 50%, concentration of extract that kills 50% of the viable cell population.

The results of the assay showed that the samples of packaging material, before and after sterilization by ethylene oxide (EtO), resulted in viability of over 90%, and therefore do not become cytotoxic.

3.4 Sterility assay and ethylene oxide residues

The samples were classified according to their position (0 to 5a; 0 to 5b) during exposure to EtO, for the penetration analysis. The outermost or surface position corresponds to the number 5, and the innermost position to the number 0. (Figure 6)

Sterility tests were carried out through the analyses of two biological indicators (bi 3M - ATTEST –™ Bacillus atrophaeus and Terragene – Bionova BT40 - Bacillus atrophaeus). Incubation time was 48 h at a temperature of 35°C ± 1.5.

Samples of packaging were also submitted to direct incubation for 7 days with TSB liquid culture at a temperature of 35°C ± 1.5. The methodology used is in accordance with the Brazilian Pharmacopoeia. The analyses of the three sterility tests (biological indicator Bionova BT40, 3Mattest and direct incubation) confirm the sterility of the packaging material after being submitted to ethylene oxide gas.

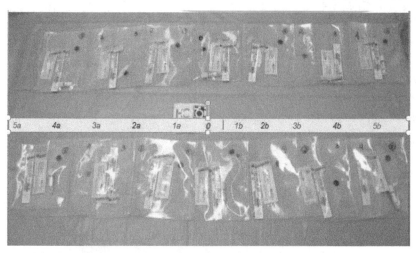

Fig. 6. Samples of packaging positioned from 0 to 5 for EtO penetration analysis.

Fig. 7 and 8. Performance of the sterility test by direct incubation in samples submitted to ethylene oxide.

The analysis of ethylene oxide residues was performed by the Gas Chromatography test, determining the levels of Ethylene chlorohydrin and Ethylene glycol. The data are shown in **Figure 9** and show that the levels are within the limits accepted by our legislation.

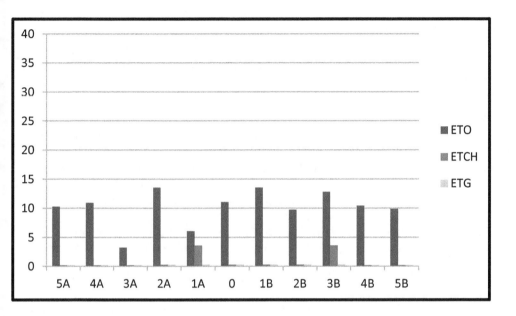

Note. Maximum limits according to the Brazilian legislation. [ETO up to 25/ ETCH up to 25/ETG up to 250]

Fig. 9. Residues (ETO, ETCH, ETG) found in the samples submitted to sterilization.

4. Final considerations

Analyses of the packaging used in this study demonstrated that it is a good option for cryopreservation of tissues at a temperature of – 80°C.

Our experience with assays to validate coextruded polyethylene and polyamide plastic film shows that the mechanical properties of this material are not altered by cryopreservation and sterilization. Penetration resistance of the thermal seal remained unaltered after all the processes carried out in a tissue bank, such as sterilization and cryopreservation.

We found a loss of barrier due to increased oxygen permeability of around 10% after sterilization and cryopreservation, which can be explained by the humidification of the polyamide. However, this slight alteration in oxygen permeability does not compromise the inner vacuum of the packaging, and does not place at risk the tissue packaged in it.

In relation to total migration, we did not observe any alterations in the assays, i.e. once again, sterilization and cryopreservation did not lead to monomer migration at levels above those required by our legislation. This is also valid for specific migrations of 1-octene, ε-caprolactam and hexamethylenediamine.

The application of ethylene oxide is safe for sterilization of this type of packaging, as it results in good penetrability and safe levels of Eto, Etch and Etg residues at the end of the procedure. In the cytotoxicity test, we observed levels of cell viability of over 90%, therefore they do not become cytotoxic.

Thus, analyses of coextruded plastic polyethylene and polyamide film used in this study proved to be a good option for cryopreservation of tissues at temperatures of –80°C, even for prolonged periods of 150 days.

5. References

American Association of Tissue Banks. Standards for Tissue Banking. 11th ed. Mclean : American Association of Tissue Banks; 2007.

ASTM INTERNATIONAL. ASTM F1927-07: standard test method for determination of oxygen gas transmission rate, permeability and permeance at controlled humidity through barrier materials using a coulometric detector. Philadelphia, 2007. 6 p.

ASTM INTERNATIONAL. F 1306-90 (2008) e1: standard test method for slow rate penetration resistance of flexible barrier films and laminates. Philadelphia, 1990. 5p.

ASTM INTERNATIONAL. F 88/F 88M-09: standard test method for seal strength of flexible barrier materials. Philadelphia, 2007. 11p.

BRASIL, Leis etc. Decreto n.2268 de 30 de junho de 1997. Dispõe sobre a remoção de órgãos, tecidos e partes do corpo humano para fins de transplantes e tratamento. *Diário Oficial da União*, Brasília (DF). 1997 30 jun; seção 1:1.

BRASIL, Leis etc. Lei n.9434 de 5 de fevereiro de 1997. Dispõe sobre a remoção de órgãos, tecidos e partes do corpo humano para fins de transplantes e tratamento. *Diário Oficial da União*, Brasília (DF). 1997 5 fev; seção 1:25.

BRASIL, Leis etc. Portaria n.1686 de 20 de setembro de 2002. Dispõe sobre a regulamentação para funcionamento de banco de tecidos músculo esqueléticos. *Diário Oficial da União*, Brasília (DF). 2002 24 jul; seção 1:1.

BRASIL, Leis etc. PORTARIA N° 2.600, DE 21 DE OUTUBRO DE 2009. Aprova o Regulamento Técnico do Sistema Nacional de Transplantes. *Diário Oficial da União*, Brasília (DF). 2009 21out.

BRASIL, Leis etc. Resolução n. 220 de 27 de dezembro de 2006. Dispõe sobre o Regulamento Técnico para o Funcionamento de Bancos de Tecidos Musculoesqueléticos e de Bancos de Pele de origem humana. *Diário Oficial da União*, Brasília (DF). 2006 29 dez.

BRASIL. Agência Nacional de Vigilância Sanitária.ANVISA. Resolução n° 105 de 19 de maio de 1999. Aprova o regulamento técnico sobre disposições gerais para embalagens e quipamentos plásticos em contato com alimentos. Diário Oficial

[da] República Federativa do Brasil, Brasília, DF, 20 maio, 1999a. N. 95, Seção 1,p.21-34.

BRASIL. Ministério da Saúde – Agência Nacional de Vigilância Sanitária. Resolução - RDC n° 51 de 26 de novembro de 2010. Dispõe sobre migração em materiais, embalagens equipamentos plásticos destinados a entrar em contato com alimentos. Diário Oficial da República Federativa do Brasil, Brasília, DF, 22 dez. 2010. Seção 1, n. 244, p. 75-79.

Cory,AH; Owen,TC; Barltrop,JA; Cory, JG. (1991). Use of an aqueous soluble tetrazolium/formazan assay for cell growth assays in culture. Cancer Comm., 3, 207-212.

European Association of Tissue Banks. Common Standards for Tissues and Cells Banking: Berlin: European Association of Tissue Banks; 2004.

EUROPEAN COMMITEE FOR STANDARDIZATION - CEN/TS 13130-16: materials and articles in contact with foodstuffs - Plastics substances subject to limitation - Part 16: Determination of caprolactam and caprolactam salt in food simulants. February 2005. 15p.

EUROPEAN COMMITEE FOR STANDARDIZATION - CEN/TS 13130-21: materials and articles in contact with foodstuffs - Plastics substances subject to limitation - Part 21: Determination of of ethylenediamine and hexamethylenediamine in food simulants. February 2005. 17p.

EUROPEAN COMMITEE FOR STANDARDIZATION - CEN/TS 13130-26: materials and articles in contact with foodstuffs - Plastics substances subject to limitation - Part 26: Determination of 1-octene and tetrahydrofuran in food simulants. February 2005. 25p.

EUROPEAN COMMITTEE FOR STANDARDIZATION. EN 1186-1: materials and articles in contact with foodstuffs - plastics. Part 1: guide to the selection of conditions and test methods for overall migration. Brussels, 2002. 49 p.

EUROPEAN COMMITTEE FOR STANDARDIZATION. EN 1186-14: materials and articles in contact with foodstuffs - plastics. Part 14: test methods for "substitute tests" for overall migration from plastics intended to come into contact with fatty foodstuffs using test media iso-octane and 95% ethanol. Brussels, 2002. 20 p.

EUROPEAN COMMITTEE FOR STANDARDIZATION. EN 1186-3: materials and articles in contact with foodstuffs - plastics Part 3: test methods for overall migration into aqueous food simulants by total immersion. Brussels, 2002. 17 p.

EUROPEAN PHARMACOPOEIA. 6. ed. Strasbourg, France: European Department for the Quality of Medicines, 2008.

Freshney, RI (2000). Culture of animal cells: a manual of basic technique. 4.ed. New York: Wiley, 577p.

GRUPO MERCADO COMÚN. Mercosur//GMC/Res. n° 32/10. Regulamento técnico Mercosur sobre migración en materiales, envases y equipamientos plásticos destinados a estar en contato con alimentos. Disponível em: <http://www.puntofocal.gov.ar/doc/r_gmc_32-10.pdf>. Acesso em: 22 ago. 2011

International Organization For Standardization. (2010), ISO 10993-5; Biological evaluation of medical devices – Part 5: Tests for cytotoxicity: in vitro methods. Switzerland
Rockville: The United States Pharmacopeial Convention, Inc., 2010.
THE UNITED STATES PHARMACOPEIAL CONVENTION. The United States Pharmacopeia. USP 33; NF 28.

Part 2

Human Assisted
Reproduction Techniques (ART)

6

The Problem of Contamination: Open *vs.* Closed *vs.* Semi-Closed Vitrification Systems

Enrique Criado Scholz

CERAM: Centre for assisted reproduction in Marbella
Spain

1. Introduction

The Development of cryopreservation techniques, the increase in demand for cryopreserved cells or tissues and the use of these techniques in cells or tissues from patients with infectious diseases, has forced us to reduce the risk of contamination during the freezing process and the risk of cross-contamination during the storage of this material. Recent publications that demonstrate the survival of pathogens at low temperatures and possible contamination of the cells or tissues stored have changed the laws of each country and the customs and protocols used so far in the cryopreservation.

To understand the problem of contamination in cryopreservation we need to have an overview of the current problem in which all researchers are concerned about, seeking a cryopreservation protocol with good results but without contamination problems. Discussing the cryopreservation's different techniques such as slow freezing, vitrification, kinetic vitrification (extra-, hyper-, super-, ultra-fast vitrification) and the various components that help us understand the difficult balance between technique, device used and the risk of contamination. We need to use new products and new protocols' to have good results ensuring biological samples and patient safety.

Dr. Katkov's idea of find the *"Universal crypreservation protocol"* (see the other Chapter in the Book) by Katkov at al. that we can use worldwide with all the possible biological samples would lower considerably the price of cryopreservation process and we would have better results because everybody would work with the same protocol and the same results.

We have to comprehend the difference between open, closed or semi-closed devices and the importance of choosing one device or another both in morphological survival post thaw cell or tissue as on non-contamination of these samples. The device used, the protocol used and the cooling solution used can change the outcome of cryopreservation and therefore we have to find a protocol for cryopreservation with a cooling solution and a secure device to provide us good results free of contamination.

2. Contamination and cross-contamination

The first thing we must learn is to differentiate their respective importance are the concepts of contamination and cross contamination of samples. The first relates to the contamination of the sample by freezing or by direct contact with the cooling solution and the second refers to the contamination of the sample within the common container which is in contact with all cryopreserved samples, some samples may be contaminated or the liquid nitrogen (LN2) might be contaminated producing a possible cross-contamination. The potential for disease transmission and pathogen survival through contaminated LN2 has been proposed by many authors (1-3), and the evidence of contamination in human patients has been described for different pathogens (4-10). It has to be stated that none of the reported infections after insemination or ET in humans and domestic animals can be clearly attributed to the applied cryopreservation and storage procedure but the use of safe cryopreservation protocol is very important to avoid human cell contamination or cross-contamination in common LN2 tanks.

Although cryopreservation had a boom in the mid 70's and early 80's with the opening of the first sperm banks in America and Europe, it was not until the mid-80's when we saw the need for biological samples cryopreserved in quarantine and the lack of screening leads to infection of several recipients that had been inseminated with semen samples from donors HIV+ those unaware of their disease (11). In these cases it was found that samples stored in the same containers with frozen HIV+ samples were not contaminated, otherwise in 1995, six patients undergoing cytotoxic treatments hermetic problems developed an outbreak of acute hepatitis B after undergoing an autologous cryopreserved material that had been stored in the same cryogenic container as other patients infected with hepatitis B (12).

2.1 Cells and tissue contamination

In the field of assisted reproduction, although it hasn't been detected any contamination in the cryopreservation of gametes and embryos, the probability and the occurrence is low, the risk is not zero so it is recommended to follow the rules in biosecurity manuals for both the physical and chemical risk as well as the risk of contamination and cross contamination of samples.

The case in 1985 where there was infection with hepatitis B in the cryopreserved samples (12) the infection was due to an error in packaging and storage of samples. With time a deterioration of the bags containing infectious material causing the infection of the LN2 and other samples was observed.

Further studies have shown that the storage of samples is decisive. There is evidence that frozen samples in hermetically sealed straws are not contaminated even if they are in contaminated containers with contaminated LN2 and LN2 does not contaminate infective biological samples that were frozen in a sealed container (13,14)

During the cryopreservation, biological samples go through many processes before being cryopreserved. In the case of IVF cells are subjected to a phase of procurement, fertilization, development, transfer and finally cryopreservation. This represents an approximate 6-day process in which many factors can affect the contamination of the sample at the end of the process. We can find contamination or cross-contamination in the following cases (15):

- Handling contaminated biological samples (semen, follicular fluid, tissue, etc.). Without precautions to avoid contamination outside the base plate to be used for conservation (cryotube, straw, etc.). It is very important to disinfect and clean the container before filling it with LN2 (16). In this regard to ensure an adequate level of biosafety a study is needed of infectious diseases transmissible from any patient or donor who wants to freeze any samples. According to Castilla (17) the clinic policy for a donor with infectious diseases is radically different to that of a patient with any of these diseases wanting to freeze biological material for autologous use. In the first situation, the biological material at hand will not freeze. In the second, the biological material should be frozen but with measures that we discuss later. Screenings for infectious diseases that normally must be submitted are: To analyse serological studies for syphilis, hepatitis and HIV. To analyse the clinical studies infective clinical phases: toxoplasmosis, rubella, herpes virus, cytomegalovirus (CMV), Neisseria gonorrhoeae and Chlamydia trachomatis. These tests are required for donors of semen every 6 months. As the risk of disease transmission during storage in LN2 is mainly viral. Interestingly, the American Society of Fertilisation (18), ESHRE (19), British Andrology Society (BSA) (20) and the Spanish Association of Tissue Banks (AEBT) (21) also recommend serologic screening for CMV, not just clinical. The presence of CMV in semen has been associated with active disease (anti-CMV IgM + or recent seroconversion anti-CMV IgG +). Similarly, these companies recommend performing serologic tests for HTLV-I and HTLV-II. But although it is clearly demonstrated the transmission of human papilloma virus by using LN2 cryotherapy and has been shown IUI transmission of herpes simplex virus (HSV) (22), none of the scientific associations mentioned above recommend a culture for detection or serological studies of HPV donors or patients with infection who are going to freeze biological material because the analysis to detect these deceases are not very sensitive. As rubella serologic screening of donors, its low prevalence in this population means that serological tests have a low positive predictive value, making it unadvisable. Finally, we believe a patient who needs to freeze some reproductive biological material should have at least one serology for HIV, hepatitis B and C. This proposal is consistent with the recommendations of the AEBT for cryopreservation of semen (21).

• Use of contaminated culture media. In these cases the degree of cross-contamination would reach very high levels having an impact on many patients. Although the preparation of embryo culture media and sperm extenders from specific ingredients are avoided in human clinics, it continues to be a common practice in animal ART (23). Nevertheless, many ingredients of embryo culture media and sperm extenders act as stabilizers for many micro-organisms at freezing temperatures (milk, serum or serum albumin, sucrose, sorbitol and other sugars). Unfortunately, the most common cryoprotectors (CPs) in applied oocyte cryopreservation and embryo (glycerol, DMSO, ethylene glycol, propylene glycol, methanol etc.) are toxic for cells. Also bacteria and viruses efficiently protect from cryoinjuries, eg Concentrations of DMSO as low as 5% enveloped viruses defend against the trauma of freezing (24). The Fact That micro-organisms survive in association with germplasm is not only important from the potential of disease transmission by embryo transfer to recipients, but also in approaches to the storage of samples for testing and health certification of embryos for international movement. On the other hand we must also bear in mind that all culture media containing antibiotics to prevent or limit survival of microorganisms.

- Conservation of contaminated material or straws cryotubes closed or sealed badly flawed causing the breakdown of the frozen straw, leaving the contaminated sample directly exposed to the LN2 tank risking contaminating the other samples. Closed systems can be sealed in many ways (thermal sealer, ultrasound sealer, radiofrequency sealer, polyvinyl alcohol powders, and solid caps). Given the sealing time and the temperature reached does not affect the cryopreserved sample, we have to ensure that the seal is airtight and that the device is built of resistant material to low temperatures of LN2 (Ionomeric resins, quartz glass capillary, Polyvinyl chloride, Polyethylene glycol tetralato, etc).

- Using contaminated LN2 during the freezing process. In this case we have proposed some solutions that we will see later.

- Poor source management of LN2 from our supplier contaminating commercial LN2 that comes to our lab in the process of manufacture or transportation and filling our containers.

- For transportation of contaminated material in containers. Storage containers should be emptied and cleaned periodically due to the risk of lost straws or small particles of contaminated material that falls to the bottom of a large container (25,26). Most of the companies of LN2 containers provide cleaning protocols. The main problem is the cleaning of transport cylinders called "dry" because the material that absorbs the LN2 in these bottles is difficult to sterilize. Bielanski (27) describes a method of disinfection of commercial dry shippers with two different types of a LN absorbent. Based on the results presented, it appears that solutions of sodium hypochlorite and ethylene oxide are equally useful for the disinfection of dry shippers constructed with a hydrophobic LN absorbent. In contrast, for dry shippers without a hydrophobic LN absorbent it is advisable to use gas only for decontamination in order sterilization to avoid their damage by liquid disinfectants.

- The air in the room. If the air that reaches the lab comes from another area that could be contaminated and there isn't a good filter. Some laboratories do not have filtration systems or positive pressure to prevent air contamination.

- Operators. If they are infected then that can lead to contamination by contact or peeling during processing of samples or the handling of cryogenic tanks. Staff must meet certain health and hygiene conditions: negative serology for HIV, HCV, HBV and vaccination against hepatitis B and other viral diseases for which there is a vaccine available. We must also have a detailed description of their jobs, tasks and responsibilities. In addition the centre must provide the worker training in freezing techniques for updating and improving procedures.

- Use of open devices. In recent times there is much talk of closed or open system and the possibility of contamination, so many countries have banned open systems and the trend is to ban the high risk of sample contamination. In a closed or semi-closed device the nitrogen of common container is never in contact with biological material frozen on the inside so cross-contamination cannot produced. In the open system, the biological material is in contact with the common nitrogen so contamination from the sample is very easy if the LN2 is contaminated or contamination of LN2 if the sample is contaminated. The latest study done by Criado and his group (28) showed 45% of contamination in an open device (Cryotop) Vs 0% of contamination in a semi-close device (Ultravit) equal and using a contaminated laboratory LN2.

2.2 Cooling solution contamination

The cooling solution plays a significant role in avoiding contamination of biological samples. It means that we will freeze the sample and we will deposit it for a long storage until thawed and used. Normally the LN2 cooling solution is the most widely used in cryopreservation and survival of pathogens at high temperatures (-196 ° C) has already been proven by many studies (1-3,27,28) cases also involved in seeing cross-contamination of human papillomavirus (14,29).

The need for better cooling rates to avoid formation of crystals in cryopreservation has resulted in the discovery and use of new cooling solutions (slush, slurry, etc.). So there are more components to consider when contamination is to be avoided. Using these new cooling solutions gives a lower temperature than the LN2 temperature and much faster transmission. The Slush nitrogen is obtained by a vacuum pump (Telstar TOP-3; Telstar S.A., Terrassa, Spain) that solidifies part of the LN2 in a few minutes. On return to normal atmospheric pressure, the nitrogen collapses, and the subcooled LN2 has solid particles in it commonly referred to as "slush" (30).The advantage of Slush nitrogen lies not only in the temperature difference with respect to LN2 (-196°C Vs -210°C) but also in the reduction of the Leiden frost effect, which is the formation of a layer of vapor around the sample when immersed in the cryogenic liquid from room temperature decreasing the cooling rate (31,32). It has not yet been demonstrated the survival or non survival of pathogens in this cooling solution of 15-20 ° C difference in LN2, this is obtained by vacuum pressure, which can lead to rupture of the cell wall of pathogens to balance internal and external pressure of these in the process of forming Slush. The 'Slurry' nitrogen is a mix of LN2 with different particles for example copper powder. At present investigations are being carried out as an alternative to LN2 to increase the cooling rate because with this cooling solution the thermal conduction is increased. Likewise, experiments are ongoing with various solutions to increase the thermal conduction and the cooling rate.

These cooling solutions "alternatives" are only used at the time of freezing the sample and once frozen, it passes to the general container that is filled with LN2, although these solutions where they freeze cool samples have to be sterile we have to ensure that the general LN2 container does not have contact with the frozen sample in order to not contaminate the sample and the LN2 if the sample is positive for any pathogen.

Retrospective studies in which commercial LN2 cryotanks were examined after 35 continuous years of service revealed various bacterial and fungal contaminations in the LN2 detritus (23). Many of the identified bacteria isolated in these studies were ubiquitous environmental micro-organisms and were rare opportunistic pathogens of low significance in producing disease in humans or animals (Table I). It should be acknowledged that some of the isolates may have been derived from laboratory contamination during semen and embryo processing for cryopreservation rather than genuinely being present within the sample. In agreement with Bielansky and Vajta the risk of contamination by human pathogens seems to be rather low. Components of the standard LN2 production system comprise a compressor, a cryogenerator and containers. From a practical point of view, the complete sterilization and maintenance of sterility in such a robust system might be a very demanding task, if possible at all. Accordingly, some ubiquitous bacterial agents can be expected in any commercially produced LN2. Nevertheless, it is an 'in and out' system and only air-borne contaminants are supposed to enter it (LN2 compressor) via air used for LN2 production. As they are not air-borne, it is unlikely that viral agents of human concern such as HIV, hepatitis and herpes viruses would enter the LN2 production system.

Sample tank	Identified microbial contamination			Years of storage	Total no. of stored samples
	Liquid nitrogen	Semen	Embryos		
Research					
Laboratory tanks					
1	Staphylococcus auriculatis	Nd	Nd	20	580
2	Bacillus licheniformis, Bacillus spp.	Stenotrophomonas maltophilia, Staphylococcus sciuri	Nd	15	840
3	CDC group IVc 2, Alcaligenes faecalis	Proteus vulgaris	Nd	8	460
4	Brevundimonas vesiculatis	E. coli	Nd	15	1350
5	Stenotrophomonas maltophilia	---	Nd	12	650
6	Staphylococcus capitis, unidentified Gram-negative rod, Comamonas acidovorans	Morganella morganii, Gemella mobilorum, Stenotrophomonas maltophilia, Citrobacter koseri	Bacillus subtilis, Ochrobactrum anthropi, Staphylococcus epidermidis	15	1200
7	Stenotrophomonas maltophilia, Comamonas testosteroni	Stenotrophomonas maltophilia, Citrobacter koseri	Stenotrophomonas maltophilia, Bacillus spp., Pseudomonas fluorescens, Acinetobacter lwoffii	18	1480
8	Bacillus pumilus, Eikenella corrodens	---	Stenotrophomonas maltophilia, Bacillus spp., unidentified Gram-negative rod	10	960
Commercial tanks					
9	Nd	Nd	---	10	58 456
10	Nd	Aspergillus spp.	---	30	138 430
11	Aspergillus spp.	Corynebacterium xerosis, Bacillus sphaericus	---	30	34 962
12	Nd	Stenotrophomonas maltophilia	---	35	150 000
13	Nd	Stenotrophomonas maltophilia	---	15	280 864
14	Aspergillus spp.	Photobacterium damsela	---	15	260 912
15	Nd	Bacillus sphericus, Corynebacterium spp., Staphylococcus sciuri	---	12	262 642
16	Stenotrophomonas maltophilia	Ralstonia pickettii	---	12	404 955

Nd, not detected; ---, not available for testing

Table 1. Microbiological contamination of embryos and semen during storage in LN2 (23)

One of the biggest discussions recently in the world of cryopreservation focuses on the importance of the sterility of LN2. As shown in Table I and in total agreement with Bielansky and Vajta and many other authors the commercial LN2 reaching our lab is not contaminated enough to cause any infection to freeze biological material. The major problem is common containers where the samples are deposited with a LN2 stored for months, years or even decades in contact with many samples, which, many clinics do not empty and do not disinfect, so it is in common containers where we can find the highest risk of contamination and cross contamination.

As a possible solution to minimize the risk of freezing biological material some systems have been proposed where we sterilize the LN2 and where we ensure that the sample is not in contact with LN2 with the use of semi-close devices or devices that are the only ones that guarantee a hermetic sealing of the device and avoid any risk of breakage of the solder thus ensuring the aseptic samples. The fact that LN2 can be quickly and safely sterilized could encourage the clinical application of human cell/tissue vitrification, both with open carriers and with closed systems. The problem is that if this device is an open device and is passed to the general container where all the other cryopreserved samples there is a huge risk of cross contamination, so it has not helped.

- LN2 Filtration: One of the solutions that have been developed is the filtration of LN2. Air Liquid has marketed CERALIN a liquid filtration system through LN2 ceramic filters. The CERALIN ON LINE consists of two elements of liquid filtration connected in series and inserted into a section of vacuum transfer line. The ceramic membrane is made from multiple layers formed into a multi-channel element. It is housed in a vacuum insulated pipe, itself installed close to the end-use point. The filter minimizes the pressure drop and avoids the vaporization of the LN2. Thus it avoids nitrogen losses. Several sizes are available, depending on the nitrogen flow. The efficiency of this equipment was investigated and proved in laboratory. The filter is located downstream of the nitrogen vessel. During operation, LN2 flows through the filter and over the ceramic membrane. The result is high-purity LN2 with a bacteria count of less than 1 CFU/L gas. Additionally, the large filtration area of the membrane and low level of contamination of LN2 means it is likely to be several decades before filter saturation.

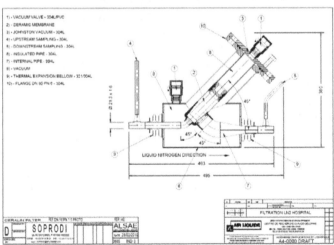

Fig. 1. CERALIN

- UV Sterilization: This method is based on emitting the minimum dose on UV radiation necessary to kill micro-organisms that can survive at the boiling point of nitrogen (-196°C) and which is irradiated in a temperature-controlled regimen, within a short time interval, before the LN2 completely evaporates. The extremely radiation-resistant bacterium Deinococcus radiodurans is inactivated (>- 4log) by administe-ring 400.000 µWs/cm² per each sterilization cycle.   An adequate amount of UV radiation de-activates the growth of all kinds of micro-organisms, from viruses like Hepatitis (which require an 8.000 UV dose) to fungi like Aspergillus Niger (330.000 UV dose) (33). At CRYO 2011 Dr. Parmegiani spoke about a new dispositive of UV sterilization of the common containers with cells or tissues inside but the scientific community thinks that is too dangerous biological samples exposed to UV rays without any protection. Although his group is proposing special canisters "not transparent" I think they have to do many more tests to rule out damage to the samples because the common view is confirmed that UV light is harmful, even if used just overnight decreased embryo developmental rates.

- LN2 Steam: As an alternative to hermetical storage in LN2, cryostorage contamination might be avoided by storing the carrier containing the vitrified oocytes in LN2 vapour (34, 35). However, Grout and Morris (36) maintain that storage in the vapour phase of LN2 still carries a risk of sample contamination. Storage of semen in LN2 vapours was discarded early in the development of sperm cryopreservation techniques and it was found that long-term viability of sperm was reduced compared with LN2 storage (37,38). However, recent experiments with new materials have succeeded in developing the technique with acceptable results for both semen and embryos (39,40) and in our last experiment we demonstrated 0% of contamination in vapor nitrogen in a experimental contaminated laboratory LN2 (non published). The drawback of the generalization of this form of storage is the need for careful monitoring of temperature in different parts of the container, which makes the marketing of these containers type (40) more difficult.

- Before entering discussions regarding the sterility in LN2 used for vitrification, we should debate the use of communal containers, which is where cross-contamination can be found, as there is a possibility that the "contaminated cells" could come into contact with each other, and where a number of viruses and bacteria are found, which would never be found in the commercial LN2.

2.3 Contamination in transport

To carry out a safe transportation of biological material we should clearly distinguish a number of concepts (17).

1. Infectious substances: those that contain viable microorganisms (bacteria, virus, prions, parasite, fungus) or bacterial toxins that are known or believed to cause disease in animals or humans.
2. Diagnostic specimens: human or animal materials (body fluids, blood, tissue, tissue fluids, etc.). Obtained for diagnostic or investigational (41).

Most often transported biological reproductive materials are cryopreserved semen donor and follicular fluid when the laboratory is separated from the follicular puncture site. In both cases, we consider the recommendations to follow are those of diagnostic specimens. There are several documents related to the transport of biological material, such as the Universal Postal Union (UPU), the International Aviation Organization (ICAO) and International Air Transport Association (IATA) (42-44).

At European level, all documents related to transport are based on the recommendations of the Committee of Experts of the United Nations Dangerous Goods (UN) (45). There is also a european agreement on international transport of dangerous goods by road (ADR), approved by RD 2115/9838 (46). We will describe some aspects of the mentioned regulations on the transport of diagnostic specimens. The basic system consists of packaging:

1. Primary container, watertight, leak proof, labeled and contains the sample. This container should be wrapped in absorbent material. In terms of labeling, according to AEBT, if it is a semen sample from a donor, must contain an alphanumeric code that identifies the donor and the sample number of the donor. On the other hand, if the sample is for autologous use may be noted also the surname of the patient (21).
2. Secondary container, sealed, leak-proof and protects the primary container. You can place multiple primary containers wrapped in a secondary container. This should be sufficient absorbent material used to protect all primary containers and avoid collisions between them.
3. Outer shipping container: the secondary container is placed in a shipping package that protects the secondary container and its contents from outside elements, such as physical damage and water.

The data forms, letters and other identifying information of the sample should be placed taped outside the secondary container. The label for submitted materials consists of:

1. Basic triple packaging.
2. Does not require signs from United Nations (UN).
3. No substances require pictogram or declaration from the sender.
4. "Biological material for clinical use" must be indicated.
5. Tag address:
   • Name, address of destination, as detailed as possible, and phone number.
   • Name, address, telephone number and contact person at the semen bank.
6. The documents included with the storage conditions and special instructions for shipping. One of the special considerations that we must have in mind when transporting a sample of semen is not breaking the cold chain, so you must use a container or LN2 as well as avoiding the possible use of dry ice.
7. Permission for import / export and declaration.
8. Label orientation.
9. Date and time of departure of Semen Bank (21).

The requirements to be met for local transport are as follows:

1. Sealed and resistant containers.
2. Threaded tubes upright (rack, tray ...).
3. Use of resistant boxes and perfect closure.
4. Secured box in the transport vehicle.
5. Appropriate Labeling.
6. Have the forms with necessary details.
7. Vehicle with kit (gloves, absorbent material, disinfectant, waste container, etc.).

You must ensure perfect coordination of transport between the sender, carrier and recipient to ensure delivery. Thus, each party involved should carry out its part perfectly and appropriately. So stand out from other actions that the sender must ensure the proper identification, packaging, labeling and documentation according to established biosafety guidelines in the "Recommendations of the Committee of Experts of the United Nations Transport of Dangerous Goods" transporting must be kept in appropriate conditions (temperature, light ...) the material from which the sender receives it until it is delivered to your destination and have the appropriate licenses to perform this type of transport, and finally, the recipient must confirm with national authorities that the material can be legally imported.

According to AEBT (21), the possibility of returning a material that hasn't been used should be avoided, as a rule, the return of the semen that has been provided by the Bank, as it will only accept the return of the displayed when you meet the following 3 conditions:

1. The sample wasn't thawed.
2. You can demonstrate the integrity of the packaging (the seals are intact).
3. The temperature of the sample was maintained throughout the transport.

3. Open device

Following recent studies of cell and tissue contamination in freezing and the recent debates regarding the sterility of LN2 in vitrification processes the devices play an important role in the asepsis of the frozen sample. There are many different types of device and of various materials but from the point of view of sterility, devices can be divided into Open, Closed and Semi-closed devices.

There is a lot of controversy and confusion about the concept of Open device. For most cryobiologists Open devices are devices that allow direct contact of the biological sample to be frozen with the cooling solution but when there is contact with the interior of the device but not with the sample to be frozen it would not be considered an open device. Once inside the common cooling containers the cooling solution enters and leaves the device keeping all frozen samples in contact.

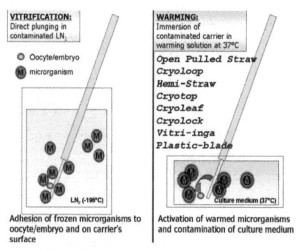

Fig. 2. Risk of Contamination with open devices (47)

Generally, using open devices the achieved cooling rates are approximately 20.000-30.000 ° C / min which favor good vitrification of the sample. The problem is that being in direct contact with the cooling solution there is a risk of pathogen transmission to the biological sample at the time of freezing and a high risk of cross contamination in the common cooling containers. They have been prohibited in many countries for this high risk of contamination and the global trend is to ban them for use with human samples. Recent microbiological studies indicate a 45% pathogen contamination (Pseudomonas and E-coli) with a simple 10-second contact with the open device (Cryotop) with contaminated cooling solution (28).

The most known and used open devices are the following:

Open pulled straw

In the OPS method, 0.25 mm standard insemination plastic mini-straws were heat-softened over a hot plate and pulled manually, as originally described by Vajta et al. (48). The inner diameter and the wall thickness of the pulled part of the straw are approximately 0.8 and 0.07 mm respectively. Cells are load into the pulled straws by placing the narrow end of the pulled straw in the third droplet of medium and aspirating oocytes within a 2–3 mm long liquid column (1–1.5 µl) using capillarity. The straws are then cooled by being plunged directly into LN2 and stored briefly. For warming, the open end of the straw is immersed vertically into 4.5 ml of the warming solution at 37°C. The solidified vitrification solution became liquid within 1–2 s. A cooling rate of 16.700°C/min is obtained with this device (49).

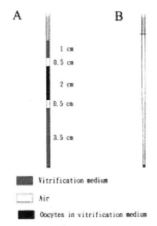

A   B

1 cm
0.5 cm
2 cm
0.5 cm
3.5 cm

■ Vitrification medium
☐ Air
■ Oocytes in vitrification medium

Fig. 3. A) The 0.25 ml conventional straw is loaded with 1 cm of vitrification medium, 0.5 cm of air, 2 cm of vitrification medium containing oocytes, 0.5 cm of air, and 3.5 cm of vitrification medium using a syringe. (B) The open pulled straw is loaded with vitrification solution (1–2 µl) containing oocytes by means of the capillary effect by a simple touch.

Cryoloop

The Cryoloop (Hampton Research, Aliso Viejo, CA, USA), used as a vessel in vitrification, is a thin nylon loop used to suspend a film of cryoprotectant containing the oocytes and directly immerse them in LN2. Vitrification of oocytes using the Cryoloop has advantages over conventional vitrification procedures in that the open system lacks a thermo insulating layer, together with the small volume of <1µl, results in both rapid and uniform heat exchange during cooling. A cooling rate of 20000°C/min is obtained with this device.

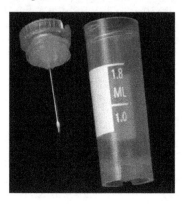

Fig. 4. Cryoloop

Hemi-straw

The Hemi-straw (Astro-Med-tec, Salzburg, Austria) is an embryo carrier that consists of a large gutter on which a small quantity of CPAs (<1µl) containing the cell is deposited. The Hemi-straw is subsequently inserted into a larger pre-cooled 0.5 ml straw (CBS, Cryo Bio System, Grenoble, France) under LN2. Prior to the commencement of the warming process the Hemi-straw is pulled out of the larger straw under LN2 and the tip of the Hemi-straw is immediately immersed into a petri dish containing a sucrose solution. A rapid cooling rate of >20.000°C/min is achieve by allowing direct contact of the biological material with LN2. (50)

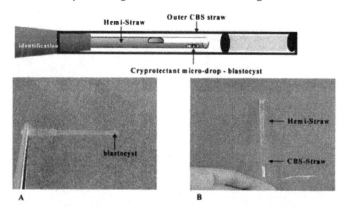

Fig. 5. Scheme of the Hemi-Straw: (A) loading the cell on the tip of the Hemi-Straw; (B) insertion of the Hemi-Straw into a larger 'CBS' straw. (51)

Cryotop

(Kitazato Supply Co, Fujinomiya, Japan) Individual oocytes were picked up in an extremely small volume (<0.1 µl) of vitrification solution and placed on top of a very fine polypropylene strip (0.4 mm wide × 20 mm long × 0.1 mm thick) attached to a hard plastic handle specially constructed according to specifications by Kitazato. The droplet volume was estimated from the length of the fluid column within the pipette tip. As soon as the oocyte was placed onto the thin polypropylene strip of the Cryotop, it was immediately submerged vertically into filtered LN2. Then, the thin strip was covered with a hard plastic cover (3 cm long) on top of the Cryotop sheet to protect it during storage in LN2 containers. For warming, the protective cover was removed from the Cryotop while it was still submerged in LN2, and the polypropylene strip of the Cryotop was immersed directly into the solution at 37°C for 1 min. A cooling rate of 23.000°C/min is obtained with this device (49).

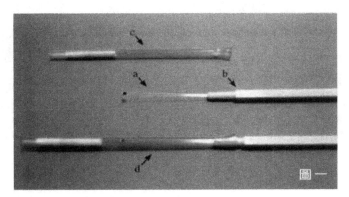

Fig. 6. Cryotop

Cryoleaf

The McGill Cryoleaf™ is very similar to Cryotop but with a number of features designed to improve the loading and storage of cells. Safety during storage has been improved, as the cells are double protected from stress and contamination through a closed cover system but not hermetically sealed leaving cells in direct contact with LN2. The McGill Cryoleaf™ and the vitrification media have been developed by Dr. Chian and Prof. Tan at McGill University, Montreal.

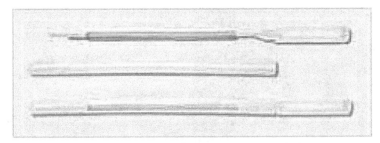

Fig. 7. Cryoleaf

Cryolock

(Biodiseño, Colombia) With this device, cells are deposited near the black mark using the minimal amount possible of vitrification solution (2 µl aprox.) The black mark eases the cover up. The Cryolock® is immediately plunged into LN2, whilst holding the Cryolock®. After this the cap is grasped with forceps and plunged into LN2 until bubbling stops, be aware to not take the Cryolock® out of the LN2 whit covered up, twist and lock gently. Finally, place the Cryolock® in the goblet with the cap downward facing and store for the desired time. For warming, remove the patients canister form the dewar and place in a styrofoam box completely cover the Cryolock® with LN2. Grasp the Cryolock® body at the indentation with forceps and remove from the goblet. Grasp the cap at the indentation with forceps, twist and pull down without taking the Cryolock® body out of the LN2, it must always remain in LN2. Finally remove the Cryolock® from the LN2 quickly and pass into thawing solution at 37°c and follow the protocol.

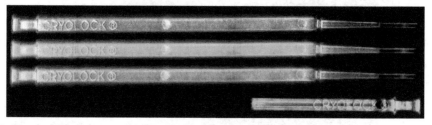

Fig. 8. Cryolock®

Vitri-Inga

The Vitri-Inga vitrification strip is an apparatus that consists of a fine, very thin polypropylene strip (0.7 mm thick) with a specially designed round tip, in which there is a minute hole to receive the cell; the strip is connected to a hard and thicker plastic handle. Vitri-Inga´ plastic sheaths are 0.5 ml semen straws with a cut in the middle. The total time from when the oocyte was placed into the vitrification solution till its immersion into LN2 is between 50 and 60 seconds. The plastic sheaths, which had been previously cooled for at least 2 min in LN2 vapor on the metal rack inside in the Vitri-Equip, are vertically immersed into LN2. The Vitri-Inga strip with the vitrified oocyte is then inserted into the plastic sheaths for storage, and transferred to a LN2 tank. (52)

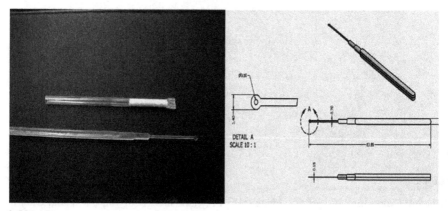

Fig. 9. Vitri-Inga

Plastic-blade

A serum Tube (Sumitomo Bakelite, Tokyo) was employed as a vessel for cryopreservation. A clear polyethylene terephthelate film (50 mm in thick) was cut into a T-shaped piece. As shown in Fig. 10, the horizontal arm of the "T" shape was rolled and fit securely to the inner wall of the cap. After equilibration with cryo-medium, the embryo for vitrification was placed at the center of a plastic blade, the vertical limb projected from the cap, and five embryos were the maximum allowed on one blade. The width of the plastic blade was significantly wider than that of the Cryotop which was commercially available tool for the storage of the vitrified human embryo. The blade was submerged directly into LN2 and inserted into the tube that was pre-cooled with LN2, then the cap was fastened on the serum tube, which accomplished preservation in the LN2 container. For warming, the serum tube that contained the plastic blade was submerged under LN2, and the serum tube was opened and the plastic blade containing vitrified cells was removed from the LN2 and placed directly into the well of the base medium at 37°C. (53)

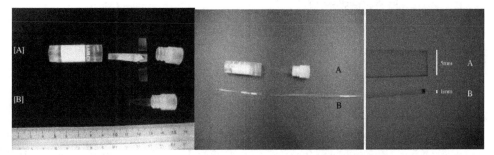

Fig. 10. Plastic-Blade

4. Closed device

Closed systems were born as a need to solve this direct contact with the open devices. The pioneer and first major proponent of close systems for freezing embryos, oocytes or sperm was F. Ostashko in 1960 as an aseptic alternative to the Cassou method. In such systems the biological sample is not in contact with the cooling solution at the time of freezing or at the time of storage in the common containers. This prevents contamination by contact and cross contamination from shared containers. The main feature is that the cooling rate is much lower with these closed devices. By lowering the cooling rate most vitrification protocols with closed systems have a high concentration of cryoprotectants to prevent crystal formation making them "dangerous" protocols for the cell due to the cytotoxicity of the cryoprotective substances. Many comparative studies of open and closed devices listing very similar results.

In the market there are many closed systems to vitrify and more appear daily due to the emphasis that cryobiologists put into to finding the perfect vitrification system that will prevent contamination of the sample and cross-contamination allowing a survival and cell viability with a protocol free or low of cryoprotective substances. The closed devices can be closed or sealed in many ways but most importantly a hermetic seal must be made, preventing entry to the inside and leakage of pathogens to the outside. Thermo seal, radiofrequency seal and ultrasound seal are some of the most used systems that ensure that the stalled sample remains suspended in time.

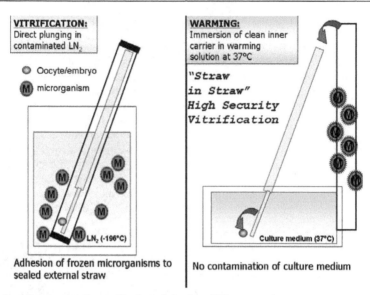

Fig. 11. Risk of Contamination with closed devices (47)

Amongst the most commonly used closed systems are the following:

25 to 0.5 ml Straw

This was one of the first devices used to freeze semen, oocytes, embryos or tissues. If the device is not hermetically sealed they are open devices, but if they are hermetically sealed they are closed systems, as the sample does not come into contact with the cooling solution. The main problem is the cooling rate, as the device is constructed from PVC or ionomeric resin and having a substantional wall thickness there is little temperature transmission (CBS). The cooling rate achieved by these straws is approximately 2.500 ° C / min (49)

Fig. 12. straw

CVM Ring Fibre Plug

CVM™ (Cryologic, Australia) involves the rapid cooling of specimens without their immersion in, or direct contact with the cooling solution. This reduces the risk of any potential contamination by pathogenic microorganism that may be present in the cooling solution. The specimens are put into a droplet which is transferred to the hook at the end of a custom designed fibre called a Fibreplug™. The Fibreplug™ is then transferred to the specially treated surface of a CVM™ Block that has been chilled to LN2 temperature. The droplet vitrifies into a glassy bead and the Fibreplug™ is placed securely into a pre-cooled CVM™ sleeve. A cooling rate of 10.000°C/min is obtained with this device. Besides the cooling rate, another main problem with this method is that to cool the CVM block the surface makes contact with LN2 thus "contaminating" the surface, which could then provoke a contamination of the sample.

Fig. 13. CVM Ring Fibre Plug

Rapid-i™

Rapid-i (Vitrolife, Sweden) is based on the same principle as the open vitrification system of the Cryoloop meaning that the embryos are place in a minute volume of vitrification solution in a hole and held there by surface tension. The Rapid-i™ holding the embryos is in turn placed in a pre-cooled RapidStraw™ sitting in the container filled with LN2. This unique feature of the Rapid-i™ vitrification System means that vitrification actually takes place in super-cooled air reducing contamination risks. The straw is sealed after vitrification making the critical time frames of the dehydration steps easier to keep and creating an aseptic vitrification system without any contact between vitrified material and LN2. The main problem is that a cooling rate of 1.200°C/min is achieved with this device (54).

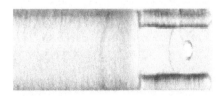

Fig. 14. Rapid-i

Vitrisafe

Is a modification of the previous Hemi-straw vitrification plug that allows a complete insertion in high security 0.5 ml straw. The Vitrisafe consists of a large gutter that is totally inserted into a larger pre-cooled 0.5 ml straw (CBS, Cryo Bio System, Grenoble, France). Only after welding both ends of the 0.5 ml straw to ensure the complete isolation of the biological sample is the complete straw plunged into LN2. For warming, the gutter is removed from the outer straw without contact with LN2 and the tip containing the biological material is directly plunged into the dilution solution in order to archive a rapid warming. A cooling rate of 1.300°C/min is archived in the vitrification process (50).

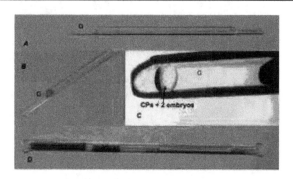

Fig. 15. Vitrisafe

High security vitrification straw

The CBS™ High Security vitrification straws are made from an ionomeric resin that is chemically inert, biocompatible and has physical characteristics resistant to ultra low temperatures and pressures created by expanding liquids and LN2. Sealed straws are resistance tested to 150 kg/cm2 (2133 lb/sq.inch), both the seals and the material should resist in order to have the batch approved. The HSV (High Security Vitrification) kit is composed of a High Security ionomeric resin straw, a capillary tube with a pre-formed gutter attached to a colored handling rod and a blue plastic insertion device. For freezing the sample is deposited into the gutter a few millimeters from the end using a micropipette. The drop holding the sample must be under 0.5 µL. immediately place the capillary rod and handler into the straw and push until the rectangular portion of the handler comes in contact with the flared end of the straw. While still holding the straw in place, seal the open end, hold the straw using tweezers in the area of the handling rod and quickly plunge the entire straw into LN2 vertically. For thawing lift the straw enough to expose the colored handling rod. Make sure the end with the sample remains immersed in the LN2. Holding the straw, use the opening device for HSV kit to section the straw and immediately (within 2 seconds), plunge the gutter into the first dilution media. A cooling rate of 2.000°C/min is archived in the vitrification process (55).

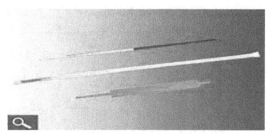

Fig. 16. High security vitrification straw

CryoTip

(Irvine Scientific) A plastic straw container which can be sealed as a closed device to hold gametes or embryos in a specialized medium during cryopreservation procedures and subsequent long term storage in a LN2 tanks. CryoTip consist of a drawn plastic straw with an ultra fine tip and a protective metal cover sleeve. This device has been optimized as a closed system for cryopreservation procedures. For freezing aseptically remove one CryoTip when ready to use. Aseptically attach the wide end of the CryoTip to an aspiration tool, such as a luer tip syringe, using the Connector. When specimens are ready to load into the CryoTip, aseptically slide the metal cover sleeve carefully along the straw to expose the fine tip end. Gently load the specimens into the CryoTip by aspiration using the plunger on the syringe to control the uptake of medium and specimens. Heat seal the fine tip below the 1st mark, then slide the metal cover sleeve down over the fine tip to protect it and plunge the sealed CryoTip into the LN2 reservoir. A cooling rate of 12.000°C/min is obtained with this device (49)

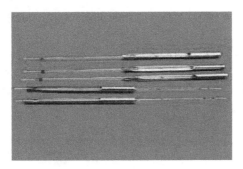

Fig. 17. Cryotip

Cryopette®

(Origio) It is derived from the original STRIPPER® family of denudation tools. It includes a sterile STRIPPER® tip with an integrated bulb to facilitate loading. This eliminates awkward external handles, rods, and pick-up tools required with other devices and guarantees simplicity, speed and ease of use. The bulb is designed to deliver the sample to the desired location every time. The maximum load volume is 1.2 µl, producing a cooling rate of approximately 23.700°C/min.

Fig. 18. Cryopette

These last two devices have been criticized by Parmegiani and his group's latest articles and in the last CRYO congress 2011, for the "potential danger" they have in his opinion: In the thawing process the external part of the device is in direct contact with the warming solution and any pathogen that could be on the exterior could pass onto the sample. In my opinion it is a very remote hypothesis and the probability that this could occur is minimal in comparison with potential contamination of an open device, a bacteriological study would be necessary that could demonstrate that with a high % of contamination, a simple contact of the end of the device with the warming solution is sufficient to contaminate the sample.

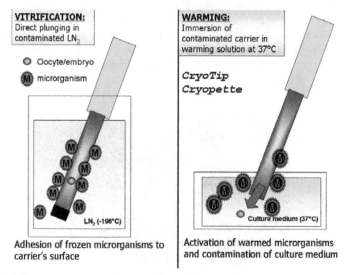

Fig. 19. Risk of Contamination with some closed devices (47)

Ultravit

Ultravit is a novel device composed of a 0.3 mm internal diameter quartz glass microcapillary tube and a flexible, transparent inert sheath that has been designed to protect and prevent it floating in the LN2. Loading the internal microcapillary tube and removing the cells from the device is very simple and easy using a syringe. Before warming, the protective sheath is cut and the internal microcapillary tube is placed in a sterile medium at 37°C after the thawing protocol. The open end of the sheath can be sealed ultrasonically in milliseconds without affecting the temperature inside the microcapillary tube, closing the system and ensuring a hermetic seal, thus preventing cross-contamination. The last microbiological control of Ultravit showed that the 5-10 seconds contact of Ultravit's internal part with contaminated LN2 (E.Coli and Psudomonas) is not sufficient to produce direct contact of cells with the cooling solution and does not result in contamination (0% of Ultravit Vs 45% of Cryotop) (28). In this study we didn't find contamination in the microdrops into which we emptied the contents of the microcapillary, also submerging the end of this (0.2mm diameter, 0.01mm of wall thickness and 1 mm in contact with the warming drop). There is a great difference between thawing with an open device (in which the entire strip is submerged in the warming solution with a surface of 42-50 mm2) and thawing with Ultravit with 95.5% less of surface in contact with the warming solution at 37°C (1.7-1.9 mm2). With Ultravit protocol, only the end of the microcapillary touches the base of the dish used to thaw, but at no time does the external part touch the warming drop (56). The following diagram shows the use of Ultravit, presuming that the cooling solution is contaminated:

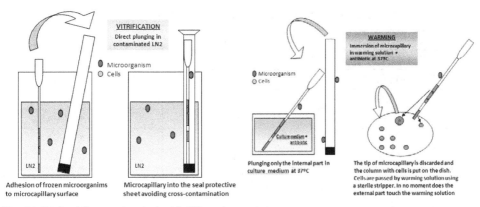

Fig. 20. Risk of Contamination with Ultravit

Our work has demonstrated that the microorganisms that may be in the cooling solution, on the outside of the microcapillary, cannot come into contact with ultra-vitrified cells inside due to the loading procedure, the contact time with the cooling solution and the diameter and surface of Ultravit making it a secure device and is enough to exclude the theoretical danger of contamination. The cooling rate obtained with Ultravit was 250.000 $C°/min$ (57) with Slush nitrogen allowing ultra-vitrification with low concentration of CPA (1.5-2 M) and a morphological survival rate of 92% of human mature oocytes and 59.1% of blastulation rate in mouse embryos.

5. Semi-closed device

As a consequence of the necessity of a device which avoids contact of the biological sample with the cooling solution but that would achieve cooling rates high enough to ensure a high rate of vitrification, semi-closed devices were designed. Gabor Vajta was one of the first to hypothesize the enclosure of open carriers (after direct contact of cells/LN2) in pre cooled hermetical containers (48). There are systems in which there is direct contact of the biological sample with the cooling solution only at the time of vitrification. Once the sample is vitrified , the device is placed in a protective sheath which is hermetically sealed before being passed to the communal container. This ensures no cross-contamination in the tanks. As stated earlier in this chapter, microbiological studies performed (23) showed that many of the Identified bacteria isolated in the commercial tanks are ubiquitous environmental micro-organisms and are rare opportunistic pathogens of low significance in producing disease in humans or animals so these devices are an important tool for high survival of biological samples to avoid cross-contamination. Theoretically all open systems can become semi-closed systems to protect the biological sample from cross-contamination with a high cooling rate but the most important is:

OPS safe method

Vajta in 1997 devised a vitrifying system with OPS, but once submerged in LN2, the OPS straw is transferred to a 0.5ml CBS straw. Using this method a cooling rate of approximately 16.700 °C/min is achieved, but once passed into the 0.5ml CBS straw it is protected from cross-contamination in the common tanks.

A summary of the cooling rates obtained with each device is as follows:

DEVICE	VOLUME	COOLING RATE
CRYOLOOP	>1 µl	20.000 °C/min
HEMI-SATRAW	>1 µl	>20.000 °C/min
CRYOLEAF	>1 µl	23.000 °C/min
VITRI-INGA	1 µl	20.000 °C/min
CVM-RING	>1 µl	10.000 °C/min
VITRISAFE	>1 µl	1.300 °C/min
HSS	0.5 µl	2.000 °C/min
0.25 ML STRAW	25 µl	2.500 °C/min
OPS	1 µl	16.700 °C/min
CRYOTOP	0.1 µl	23.000 °C/min
CRYOTIP	1 µl	12.000 °C/min
RAPID-I	0.5 µl	.1.200 °C/min
CRYOPETTE	1.2µl	23.700 °C/min
ULTRAVIT	0.2 µl	250.000 °C/min

Table 2. Different cooling rates

6. Cooling rate Vs closed systems

Today the differences between Slow freezing and Vitrification are known worldwide. We all know that slow freezing is characterized by a prolonged cooling curve and the use of low concentrations of cryo-protectors generally "non-toxic" for the cells (1-2 M) with cell injury due to ice formation (58) and that Vitrification is characterized by the rapid procedure and the use of a high concentration of cryo-protectors (4-6M) to prevent cell damage that is toxic to most mammalian cells (59-67).Thus, vitrification with a semi-close devices have a better cooling rate without cross-contamination or novel cyopreservation techniques are needed that allow rapid cooling to achieve vitrification in the absence of high concentration of CPA or if is possible without CPA.

The requirements and relationships for conditions to achieve satisfactory vitrification in the area of mammalian ART are well displayed in the equation of Yavin and Arav (68)

$$\text{Probability of vitrification} = \frac{\text{Cooling and warming rates} \times \text{Viscosity (CPA concentration)}}{\text{Volume}}$$

Fig. 21. Probability of vitrification by Yavin and Arav (68)

The main points to be gathered from this relationship are that the smaller volume of the vitrification solution in which the cellular material is placed for the vitrification process, the faster cooling and warming rate that can be achieved and the lower concentration of CPAs needed reducing the detrimental effect of the inherent toxicity of CPAs and increasing the overall success of the procedure (50).

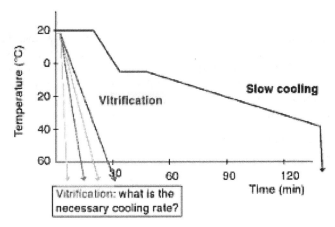

Fig. 22. necessary cooling rate to have a good probability of vitrification

What would happen if We could vitrify without CPA's or with a low concentration of CPA's? What would happen if We could combine the advantages of Slow freezing and Vitrification and vitrify with low concentrations of CPA's with a secure and free contamination device? That is Kinetic vitrification (Ultra-vitrification). Perfecting the techniques of Vitrification has been achieved a morphological survival rate comparable to normal Vitrification protocol (30) or a 59.1% of blastulation rate in mouse embryo (69) with Kinetic vitrification and concentrations of CPA's typical of Slow freezing.

Previous studies have tried to achieve high cooling rates for cell vitrification. However, none of them utilized low CPA concentrations (1.5-2 M). In 1985, Rall and Fahy successfully vitrified mouse embryos in 6.5 M cryoprotectant cocktail solution (70). In that case the method consisted in a 0.25 ml straw container plunged into LN2; the cooling rate was 2.500 °C/min. When this container was plunged into Slush nitrogen, the cooling rate increased up to 4000 °C/min (71). The use of OPS (instead of the 0.25 ml straw) in LN2 increases this cooling rate up to 5.300 °C/min (71) and to 10.000–20.000 °C/min if plunged in Slush nitrogen (71,72). Similar cooling rates were achieved in the case of a Cryoloop quenched in Slush nitrogen (73). The use of electron microscope copper grids has also been investigated, but the cooling rates were in the same order of magnitude that the afore mentioned works: 11.000–14.000 °C/min in the case of plunging the grid in LN2 (74) and 24.000–30.000 °C/min if plunged in Slush nitrogen (74,75). From Boutron's theory, none of these approaches reaches the critical cooling rate to achieve vitrification with low concentration of CPA (1.5-2M). It's impossible to use open devices with Slush nitrogen as the cell is on the outside and there is a possibility of detaching from the device.

Adjusting to Yavin and Arav formula the Ultra-vitrification technique arose achieving a cooling rate above 250.000 °C/min and of 90.000 °C/min in thawing. This rate is one order of magnitude higher than the highest cooling rate achieved in different strategies (electron microscope copper grids in Slush nitrogen (74,75), whilst keeping all the advantages of a straw-like form for the container and being in the range of the necessary cooling rate to achieve vitrification. To have this increase in the cooling rate a few changes were made to the normal vitrification process:

Slush Nitrogen

As a cooling agent this technique uses Slush nitrogen, much colder than LN2 (-196°c Vs -210°C) and with the property of avoiding the Leiderfrost Effect. When something is submerged in LN2, bubbles rise to the surface through the device, varying the thermal conductivity from the outside into the inside of the device. This does not happen with Slush nitrogen. Slush nitrogen is achieved with a vacuum pump in 5 to 10 minutes and it remains slush for a further 5 – 10 minutes before returning to liquid.

It was shown for oocytes and embryos that increasing the cooling rate would improve survival rates by up to 37% (76)

Model	Survival slush (%)	Survival LN (%)	Sig.	Publication
Bovine MII	48	28	$P<0.05$	Arav & Zeron (1997)
Ovine GV	25	5	$P<0.05$	Isachenko et al. (2001)
Porcine blastocysts	83	62	$P<0.05$	Beebe et al. (2005)
Bovine MII	48	39	$P<0.05$	Santos et al. (2006)
Mouse four-cell embryos with biopsied blastomere	87	50	$P<0.05$	Lee et al. (2007)
Rabbit embryos	92	83	NS	Papis et al. (2009)
Porcine blastocysts	89	93	NS	Cuello et al. (2004)
Mouse MII	>80	>80	NS	Seki & Mazur (2009)
Rabbit oocytes	82	83	NS	Cai et al. (2005)

LN, liquid nitrogen; GV, germinal vesicle; Sig., statistical significance; NS, not significant.

Table 3. Effect of cooling rate on survival; comparison between LN2 and Slush nitrogen (75)

Quartz Micro-capillary

Another determining factor to achieve a high cooling rate is the device used. To increase the thermal conductivity and minimize the volume, this technique has used a quartz micro-capillary. This has a 0.2-0.3 diameter allowing to ultra-vitrify 0.1-0.2 μl with a 0.01mm wall, a lot thinner than any other device (0.075 mm in OPS). Another important characteristic is the material it is made from: Quartz. The thermal conductivity of quartz glass is a lot higher than that of plastic of which other devices are made of. This converts it in one of the materials that best conducts the temperature (77)

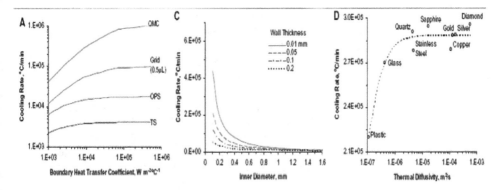

Fig. 23. Xiaoming He et al (77)

In a thermal performance of quartz capillaries for vitrification done by Risco and his group (57) a commercially available version of the OPS (MTG Medical Technological Vertriebs, GmbH) was used. The thermal conductivity of these PVC straws was 0.19 W m_1 K_1. The inner diameter is 0.800 mm and the thickness of its wall is 0.075 mm (Fig. 1a). The QC used (The Charles Supper Company, Inc.) have an inner diameter of 0.180 mm and a wall thickness of 0.010 mm. These geometrical improvements (4.44 times smaller in diameter and 7.50 times thinner) translate not only into a faster heat transfer, but also into a 20 times reduction in volume of the contained solution (for a given height). This is beneficial because the thermal conductivity of the quartz glass is 1.3 W m_1 K_1, that is almost one order of magnitude higher than that of PVC.

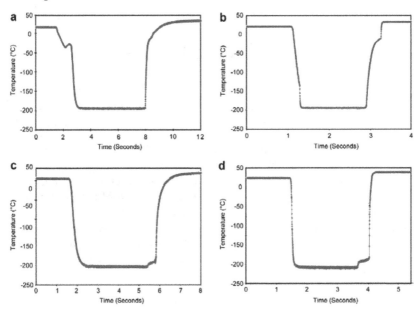

Fig. 24. (a) Thermal history for the OPS when filled with a 1.5M propane-1,2-diol and 0.3M sucrose cryoprotectant solution quenched in LN2 and then thawed in a water bath at 37°C.

A clear heat release peak is present during cooling as well as melting during rewarming. (b) Thermal history for QC when filled with a 1.5M propane-1,2-diol and 0.3M sucrose cryoprotectant solution quenched in LN2 and then thawed in a water bath at 37°C. Crystallization of water is not obvious during cooling, but melting is shown during rewarming. (c) Thermal history for OPS when filled with a 1.5M propane-1,2-diol and 0.3 M sucrose cryoprotectant solution quenched in Slush nitrogen and then thawed in a water bath at 37°C. In this case, crystallization during cooling and melting during rewarming was not recorded. However, visual inspection reveals the presence of ice. (d) Thermal history for QC when filled with a 1.5Mpropane-1,2-diol and 0.3 Msucrose cryoprotectant solution quenched in Slush nitrogen and then thawed in a water bath at 37°C. The sample keeps its transparency over all the cooling–rewarming cycle, an indication of the capability of this approach to vitrify the studied solution.

All these changes have allowed us to maintain a concentration of cryoprotectors typical of slow freezing, 2 M PrOH+0.5 M sucrose, obtaining a morphological survival rate of 92 % in human oocytes (31). Dr. Ho-Joon Lee et al (69) tested this new technique on mouse ooctyes and they saw that using Ultra-vitrification with low concentrations of cryoprotectors improved the fertilization rate and above the blastulation rate. Only the use of Ultravit device in this technique ensures the non contamination of the sample or cross-contamination in communal containers.

%	Slow Freezing (78)	Vitrification (78)	Ultra-vitrification mourine oocytes (69)	Ultra-vitrification human oocytes (31)
Surv. rate	61	91,8	92.5	92
Fert. rate	61,3	67,9	75	?
Blast. rate	12	33.1	59.1	?

Table 4. Comparison between slow Freezing, Vitrification and Ultravitrification (78, 69, 31)

This comparison demonstrates the use of low concentration of cryoprotectant in the Ultra-vitrification protocol favours the morphological survival (92%) and increases the blastulation rate (59.1%). Thus confirming the hypothesis that cryoprotectants are toxic to the biological sample and if we could find a vitrification protocol that would allow us to vitrify without cryoprotectant, we would achieve a better embryo development and a greater chance of pregnancy in the case of freezing eggs or embryos. A lot more studying is needed regarding this new technique but a priori the results indicate that we can hopefully lower the concentration of the cryoprotectants decreasing the toxicity in cells.

7. Conclusion

Cryopreservation protocols have improved and resulted in a much higher efficiency in outcomes in the last years. However, it remains important to always seek for amelioration on cryopreservation protocols and devices to ensure a major benefit and patients' safety during procedures. We need to find a method that combines a high cooling and warming rate, high survival and function of cells and tissues and is made in a way that ensures patient safety.

We must be clear that survival after a vitrification process is a "morphological survival" and that this cell has to fertilize perfectly and develop normally to rule out any damage in the process of cryopreservation and to consider a real survival.

With so much variety of devices and many different protocols, laboratories have to find the protocol and the device that best suits their skills, provided they ensure the sterility of vitrified samples and prevent cross-contamination in general containers. It has created an exaggerated paranoia about the vitrification cell and tissue contamination of the cooling solution that everything can contaminate our samples. In my opinion we must think scientifically, leaving aside the commercial interests of many of us and worrying about more logical things and research data: the probability of contamination must be demonstrated with % of contamination and leave a little aside science fiction and theoretical assumptions. For many cryobiologists avoiding cross-contamination of samples in the general containers is the most important matter as it has been proven that commercial LN2 does not have a high enough risk to have to sterilize it. The use of a closed or semi-closed device that would allow us a high cooling rate and a sealing prior to being deposited in the general container should be enough to ensure sterility and good survival and development results.

We must centre all of our interest in more practical things like finding a vitrification protocol that allows us to a vitrify without cryoprotectant, discovering new cooling solutions, discovering new materials, new devices, new procedures in which we can safely freeze samples without cryoprotectants toxicity problems, thus ensuring a good development cell after thawing, all in secure systems which ensure sterility of the sample and avoid cross-contamination. Finding an "Universal Protocol" risen several times by Dr. Katkov (79), allowing us to use the same protocol worldwide, as movement of frozen specimens around the world has increased dramatically and the lack of component preparation in the laboratory that is to thaw the frozen sample with a protocol and a device different to those they know, would not give us 100% guarantee of sample survival. Today freezing has become a luxury and not all patients can afford the excessive cost of the freezing products, if we could find a protocol without the need of cryoprotectants or a universal protocol that all could use, the price of products would decrease and the patients would benefit economically.

8. Acknowledgment

The author thanks Ms Ana Yus and Ms Cristina Gonzalez for the critical and technical support.

9. References

[1] Karrow AM, Critser JK. Reproductive tissue banking: scientific principles. San Diego: Academic Press, 1997.
[2] Vajta G, Nagy ZP. Are programmable freezers still needed in the embryo laboratory? Review on vitrification. Reprod Biomed Online 2006;12:779–96.

[3] Kuwayama M. Highly efficient vitrification for cryopreservation of human oocytes and embryos: the Cryotop method. Theriogenology 2007;67:73–80.

[4] Kleegman SJ. Therapeutic donor insemination. Conn Med 1967;31:705–13.

[5] Barwin BN. Transmission of Ureaplasma urealyticum by artificial insemination by donor. Fertil Steril 1984;41:326–7.

[6] Stewart GJ, Tyler JP, Cunningham AL, Barr JA, Driscoll GL, Gold J, et al. Transmission of human Tcell lymphotropic virus type III (HTLV-III) by artificial insemination by donor. Lancet 1985;2:581–5.

[7] Berry WR, Gottesfeld RL, Alter HJ, Vierling JM. Transmission of hepatitis B virus by artificial insemination. JAMA 1987;257:1079–81.

[8] Mascola L, GuinanME. Semen donors as the source of sexually transmitted diseases in artificially inseminated women: the saga unfolds. JAMA 1987;257:1093–4.

[9] Araneta MR, Mascola L, Eller A, O'Neil L, Ginsberg MM, Bursaw M, et al. HIV transmission through donor artificial insemination. JAMA 1995; 273:854–8.

[10] Englert Y, Lesage B, Van Vooren JP, Liesnard C, Place I, Vannin AS, et al. Medically assisted reproduction in the presence of chronic viral diseases. Hum Reprod Update 2004;10:149–62.

[11] Stewart GT, Tiler GP, Cunningham AL, Barr JA, Driscoll GL, Gold J, et al. Transmision of human T cell lymphotropic virus type III (HTLV-III) by artificial ibsemination donor. Lancet a985; 2:581-5

[12] 12 Tedder RS, Zuckerman MA, Goldstone AH, Hawkins AF, Fielding A, Briggs EM, et al. Hepatitis B transmission from contaminated cryopreservation tank. Lancet 1995; 346:137-40.

[13] Bielanski A, NadinDavis S, Sapp T, Lutze-Wallace C. Viral contamination of embryos cryopreserved in liquid nitrogen. Cryobiology 2000;40:110–6.

[14] Bielanski A, Bergeron H, Lau PC, Devenish J. Microbial contamination of embryos and semen during long term banking in liquid nitrogen. Cryobiology 2003;46:146–52.

[15] Clara González, Montserrat Boada. Bio-seguridad en el laboratorio de criobiología y aspectos legales de la criopreservación. Cuadernos de medicina reproductiva 2008.

[16] Russell PH, Lyaruu VH, Millar JD, Curry MR, Watson PF. The potencial transmission of infectious agents by semen packaging during storage for artificial insemination. Anim Reprod Sci 1997;47:337-42.

[17] Castilla JA, Magan R, Martinez L, Fernandez A, Ramirez JP, Yoldi A, Vergara F. Seguridad biologica en crioconservacion y transporte de material biológico reproductivo. Aula de formación en embriología clínica N°4 2003.

[18] American Society of Reproductive Medicine. 2002 Guidelines for gamete and embryo donation. Fertil Steril 2002;77 suppl5.

[19] Barratt C, Englert Y, Gottlieb C, Jouannet P. Gamete donation guidelines. The Corsendonk consensus document for the European Union. Hum Reprod 1998;13 supplement 2:7-9.

[20] British Andrology Society. British Andrology Society guidelines for the screening of semen donors for donor insemination (1999). Hum Reprod 1999; 14:1823-1826.

[21] Miralles A, Veiga A, Bassas L, Castilla JA, Mirabet V, Villalba R. Semen. En

[22] Estándares de la Asociación Española de Bancos de Tejidos (2ª ed), 2002.

[23] Moore DE, Ashley RL, Zarutskie PW et al. Transmisión of genital herpes by donor insemination. J Am Med Assoc 1989;261:3441-3443.

[24] Bielanski A and Vajta G. Risk of contamination of germplasm during cryopreservation and cryobanking in IVF units. Human reproduction 2009;24:2457-2467.

[25] Wallis C, Melnik JL. Stabilization of enveloped viruses by dimethyl sulfoxide. J virol 1968; 2:953-954.

[26] Steyaert SR, Leroux-Roels GG, Dhont M. Infections in IVF: review and guidelines. Hum Reprod 2000;6:432-41.

[27] Hargreave TB, Ghosh C. The impact of HIV on a fertility problems clinic. J Reprod Inmunol 1998;41:261-70.

[28] Bielansky A. Experimental microbial contamination and disinfection of dry (vapour) shipper dewars designed for short-term storage and transportation of cryopreserved germplasm and other biological specimens. Theriogenology 2005;63:1946-1957.

[29] Criado E, Moalli F, Polentarutti N, Albani E, Morreale G, Menduni F, et al. Experimental contamination assessment of a novel closed ultravitrification device. Fertil Steril 2011;95(5):1777-9.

[30] Charles CR, Sire DJ. Transmision of papovavirus by cryoherapy applicator. JAMA 1971;218:1435.

[31] Criado E, Albani E, Novara PV, Smeraldi A, Cesana A, Parini V, et al. Human oocyte ultravitrification with a low concentration of cryoprotectants by ultrafast cooling: a new protocol. Fertl Steril 2011;95:1101-1103.

[32] Cowley CW, Timson WJ, Sawdye JA. Ultra rapid cooling techniques in the freezing of biological materials. Biodynamica 1961;8:317–29.

[33] Steponkus PL, Myers SP, Lynch DV, Gardner L, Bronstheyn V, Leibo SP, et al. Cryopreservation of Drosophila melanogaster embryos. Nature 1990;345:170–2.

[34] Parmegiani L., Cognigni G.E., Filicori M. Ultra-violet sterilization of liquid nitrogen prior to vitrification. Hum. Reprod 2009;24,2969.

[35] Eum J.H., Park J.K., Lee W.S., Cha K.R., Yoon T.K., Lee D.R. Long-term liquid nitrogen vapor storage of mouse embryos cryopreserved using vitrification or slow cooling. Fertil Steril 2009. 91,1928–1932.

[36] Cobo A., Romero J.L., Perez S., de Los Santos M.J., Meseguer M., Remohi J. Storage of human oocytes in the vapor phase of nitrogen. Fertil. Steril 2010;94:1903-1907. 36. Grout B.W., Morris G.J. Contaminated liquid nitrogen vapour as a risk factor in pathogen transfer. Theriogenology 2009;71:1079–1082.

[37] Sawada Y. The preservation of human semen by deep freezing. Int J Fertil 1964;9:525-32.

[38] Ackerman DR. Damage to human spermatozoa during storage at warming temperatures. Int j Fertil 1968;13:220-5.

[39] Tomlinson M, Sakkas D. Is a review of standards procedures for cryopreservation needed? Safe and effective cryopreservation-should sperm banks and fertility centres move toward storage in nitrogen vapour?. Human Reprod 2000;15:2460-3.

[40] Clarke GN. Sperm cryopreservation: is there a significant risk of cross-contamination?. Hum Reprod 1999;14:2941-3.

[41] Calero S. Transporte material biologic. Curso Trabajar con seguridad en el laboratorio de microbiología. Aula científica 2003.

[42] The Universal Postal Union (UPU): "Manual of the Universal Postal Convention". Lays down detailed regulation for the transport of biological substances by post/mail.1995.

[43] The International Civil Aviation Organization (ICAO). "ICAO technical instructions". Safe transport of dangerous goods, 1984. Last ed:1999.

[44] The International Air Transport Association (IATA). "Dangerous goods regulations (40^a ed.)". 1999.

[45] The United Nations Committee of Expert on the Transport of Dangerous Goods (UN ECOSOC). Recommendations on the transport of dangerous goods (10^a ed). 1997

[46] Loza E, Alomar P, Bernal A, Harto A, Perez JL, Picazo JJ, et al. Seguridad en el laboratorio de Microbiología Clínica. En procedimientos de Microbiología Clínica, recomendaciones de la sociedad Española de Enfermedades Infecciosas y Microbiologia Clínica. Picazzo JJ(ed) 2000;10:1-42.

[47] L Parmegiani, GE Cognigni, S Bernardi, S Cuomo, W Ciampaglia, FE Infante, C Tabarelli de Fatis, A Arnone, AM Maccarini, M Filicori. Efficiency of aseptic open vitrification and hermetical cryostorage of human oocytes. Reproductive BioMedicine Online (2011) 23, 505– 512.

[48] Vajta G., Holm P., Kuwayama M. et al. Open pulled straw (OPS) vitrification: a new way to reduce cryoinjuries of bovine ova and embryos. Mol Reprod 1998. Dev., 51, 53–58.

[49] Shee-Uan Chen, Yu-Shih Yang. Slow Freezing or Vitrification of Oocytes: Their Effects on Survival and Meiotic Spindles, and the Time Schedule for Clinical Practice. Taiwanese Journal of Obstetrics and Gynecology 2009;48:15-22.

[50] Ri-Cheng Chian and Patrick Quinn. Cryopreservation. Cambridge University Press 2010.107.

[51] Vanderzwalmen P, Bertin G, Debauche Ch, Standaert V, Bollen N, van Roosendaal E, Vandervorst M, Schoysman R, Zech N. Vitrification of human blastocysts with the Hemi-Straw carrier: application of assisted hatching after thawing. Hum Reprod. 2003;18(7):1504-11.

[52] Fachini F, Almodin CG, Minghetti-Cámara VC, Fernandes Moron A, Nakano R, Shimabukuru L. Vitri-Inga. A new vitrification protocol. SBRA 2008.

[53] Sugiyama R, Nakagawa K, Shirai A, Sugiyama R, Nishi Y, Kuribayashi Y, et al. Clinical outcomes resulting from the transfer of vitrified human embryos using a new device for cryopreservation (plastic blade). J Assist Reprod Genet 2010;27(4):161-167.

[54] Larman MG and Gardner DK. Vitrification of mouse embryos with super-cooled air. Fertil Steril 2011;95(4):1462-6.

[55] Vanderzwalmen P. Oral presentation ESHRE 2010.

[56] Criado E. Autor reply about Contamination of single-straw carrier for vitrification. Fertl Steril 2011;95(8):e69;author reply e70.

[57] Risco R, Elmoazzen H, Doughty M, He X, Toner M. Thermal performance of quartz capillaries for vitrification. Cryobiology 2007;55:222-9.

[58] Mazur P. Freezing of living cells: mechanisms and implications. Am J Physiol 1984;247:125-42.

[59] Fahy GM, Levy DI, Ali SE. Some emerging principles underlying the physical properties biological actions and utility of vitrification solutions. Cryobiology 1987;24:196-213.

[60] Fahy GM, Wowk B, Wu J, Paynter S. Improved vitrification solutions based on the predictability of vitrification solution toxicity. Cryobiology 2004;48:22-5.

[61] Fahy GM,Wowk B,Wu J, Phan J, Rasch C, Chang A, et al. Cryopreservation of organs by vitrification: perspectives and recent advances. Cryobiology 2004;48:157-78.

[62] Fowler A, Toner M. Cryo-injury and biopreservation. Ann NYAcad Sci 2005;1066:119-35.

[63] Heng BC, Kuleshova LL, Bested SM, Liu H, Cao T. The cryopreservation of human embryonic stem cells. Biotechnol Appl Biochem 2005;41:97-104.

[64] Hunt CJ, Pegg DE, Armitage SE. Optimising cryopreservation protocols for haematopoietic progenitor cells: a methodological approach for umbilical cord blood. Cryo Letters 2006;27:73-83.

[65] Jain JK, Paulson RJ. Oocyte cryopreservation. Fertil Steril 2006;86:1037-46.

[66] Pegg DE. The role of vitrification techniques of cryopreservation in reproductive medicine. Hum Fertil 2005;8:231-9.

[67] Rall WF, Fahy GM. Ice-free cryopreservation of mouse embryos at -196 degrees C by vitrification. Nature 1985;313:573-5.

[68] Yavin S and Arav A. Measurement of essential physical properties of vitrification solutions. Theriogenology 2007;67:81-89

[69] Lee H, Elmoazzen H, Wright D, Biggers J, Rueda BR, Heo YS, et al. Ultra-rapid vitrification of mouse oocytes in low cryoprotectant concentrations. Reprod Biomed Online 2010;20:201-8.

[70] Rall W.F., Fahy G.M. Ice-free cryopreservation of mouse embryos at -196 degrees C by vitrification. Nature 1985;313:573-575.

[71] Nowshari M.A., Bren G. Effect of freezing rate and exposure time to cryoprotectant on the development of mouse pronuclear stage embryos, Hum. Reprod. 15 (2001) 2368-2373.

[72] Vajta G., Holm P., Greve T., Callesen H. Vitrification of porcine embryos using the Open Pulled Straw (OPS) method, Acta Vet. Scand. 38 (1997) 349-352.

[73] Mukaida T., Nakamura S., Tomiyama T., Wada S., Kasai M., Takahashi K. Successful birth after transfer of vitrified human blastocysts with us of a cryoloop containerless technique, Fertil. Steril. 76 (2001) 618-620.

[74] A. Arav, Y. Zeron, A. Ocheretny, A new device and method for vitrification increases the cooling rate and allows successful cryopreservation of bovine oocytes, Theriogenology 53 (2000) 248.

[75] Steponkus P.L., Myers S.P., Lynch D.V., et al. Cryopreservation of Drosophila melanogaster embryos, Nature 345 (1990) 170-172.

[76] Saragusty J and Arav A. Current progress in oocyte and embryo cryopreservation by slow freezing and vitrification. Society for Reproduction and Fertility 2011;1470-1626.

[77] He X, Park EYH, Fowler A, Yarmush ML, Toner M. Vitrification by ultra-fast cooling at a low concentration of cryoprotectants in a quartz micro-capillary: a study using murine embryonic stem cells. Cryobiology 2008;56:223-32.

[78] Yun-Xia Cao, Qiong Xing, Li Li, Lin Cong, Zhi-Guo Zhang, Zhao-Lian Wei et al. Comparison of survival and embryonic development in human oocytes cryopreserved by slow-freezing and vitrification. Ferti and Steril Volume 92, Issue 4.2009.1306-1311

[79] Katkov I.I. Race for the pace: is the universal cryoprotocol a dream or reality?. Cryobiology 2011:61;374-375.

Part 3

Farm / Pet / Laboratory Animal ART

7

Cryopreservation of Boar Spermatozoa: An Important Role of Antioxidants

Kampon Kaeoket

*Faculty of Veterinary Science, Semen laboratory, Department of
Clinical Science and Public Health, Mahidol University
Thailand*

1. Introduction

Artificial insemination (AI) is one of the first reproductive biotechnologies has been established and developed in the pig production system. In most case, liquid stored semen or fresh semen is used for AI in commercial swine herds (Wagner and Thibier, 2000). The use of FT boar semen for AI is limited due to the low fertility outcomes compared to extended fresh semen (Johnson et al., 2000; Wagner and Thibier, 2000). The first success of boar semen cryopreservation was reported in 1956 (Polge, 1956) and the first pregnancy was achieved with FT boar semen using surgical insemination in 1970 (Polge, 1970). Currently, the attempt to develop the boar semen cryopreservation technique is ongoing. Nevertheless, the success of boar semen cryopreservation is relatively variable because the factors responsible for the cryosurvival of boar spermatozoa have not been entirely elucidated.

Cryopreservation of boar semen is useful for preservation of genetic resources, improve the genetic progress and enhance the transportation of genetic material across countries (Almlid and Hofmo, 1996; Johnson, 1998). In addition, the frozen-thawed (FT) boar semen is also used with other reproductive technologies, such as in vitro fertilization (IVF), embryo transfer (ET) and sex pre-selection (Gerrits et al., 2005). Unfortunately, the advancement of sperm cryopreservation in pigs is slow, partly due to the pig producer is satisfied with the liquid stored semen and low conception rate and litter size remain the major problems when using FT boar semen (Eriksson et al., 2002.). Under field conditions, low fertility is still obtained even using FT boar semen with a sufficient motility and number of spermatozoa for insemination (Johnson et al., 2000, Eriksson et al., 2002).

The use of frozen-thawed (FT) boar semen has been developed for artificial insemination (AI) in pig long time ago in Europe and USA (Larsson and Einarsson, 1976). In Thailand, few studies on boar semen cryopreservation have been established (Buranaamnuay et al., 2006 [a,b]). However, a great variation on the survival rate of post-thawed spermatozoa are obtained, due to the lack of biological background concerning the cryopreservation technique (Buranaamnuay et al., 2006 [a,b]). During the recent years, studies on FT boar semen have dramatically improved boar semen cryopreservation technique, for instance, optimum freezing protocols (Eriksson and Rodrigez-Martinez, 2000), types of freezing package (Bwanga et al., 1991;Berger and Fisherleitner, 1992; Bwanga et al., 1991; Eriksson and Rodriguez-Martinez, 2000), semen centrifugation methods (Carvajal et al., 2004), thawing process (Eriksson and Rodrigrez-Martinez, 2000; Córdova-Izquierdo et al., 2006) and the supplement of some additives to the semen extender (Peña et al., 2003, Gadea et al., 2004; Roca et al., 2004, 2005).

Boar semen differs in several aspects from the semen of other domestic animals, for instance, the semen is produced in a large volume and highly sensitive to cold shock, the viability of the sperm cells is dramatically reduced when expose to temperatures below 15 °C (Gilmore et al., 1996). Therefore, the manipulation of boar semen requires special consideration during cryopreservation process (Johnson et al., 2000). Many factors that should be concerned for the boar semen cryopreservation included composition of diluents, type and concentration of cryoprotective agent, equilibration time, cooling rate and thawing procedure.

The relatively low fertility of FT boar semen is associated with many factors including a highly sensitive plasma membrane of boar spermatozoa against the changing in temperature during cooling, freezing and thawing process (Holt, 2000; Watson, 2000). This problem is related to the lipid composition of the sperm plasma membrane. The plasma membrane of the boar spermatozoa contains a high level of polyunsaturated fatty acids (PUFAs) i.e., docosapentaenoic acid (DPA) and docosahexaenoic acid (DHA), and had a low cholesterol to phospholipids ratio. DPA and DHA are dominant fatty acids in the plasma membrane of boar spermatozoa (Johnson et al., 1969).

During cryopreservation, PUFAs decrease dramatically due to lipid peroxidation. This is initiated when the spermatozoa is attacked by reactive oxygen species (ROS) (De Lamirande and Gagnon, 1992; Sikka et al., 1995). In mammals, the major sources of ROS formation include leucocyte, defective and dead spermatozoa (Aitken et al., 1994; Silva, 2006). The excessive ROS formation influence sperm motility, mid-piece abnormalities and sperm-oocyte fusion (Chatterjee et al., 2001; Agarwal et al., 2005).

The supplement of antioxidant compounds and some fatty acid to the semen extender, to minimize ROS formation and protect the plasma membrane function, have been used in many species (Peña et al., 2003; Gadea et al., 2004; Roca et al., 2004; 2005; Maldjian et al., 2005). It has been demonstrated that the proportion of DHA was significantly higher in the semen diluted with an extender supplemented with n-3 enriched hen egg yolks compared with the semen diluted with normal hen egg yolks (Maldjian et al., 2005). However, no study has been demonstrated clearly whether or not the supplement of DHA could improve the quality of the boar spermatozoa after cryopresevation. Rooke et al. (2001) found that DHA supplement in the boar feed increase progressive motility and normal acrosome and decrease abnormal spermatozoa. Recently, Kaeoket et al. (2008) reported that the supplement of DHA-enriched fish oil improved the FT boar semen quality. It has been shown that the supplement of cryoprotective agents (e.g., glycerol and Equex®), cholesterol analogue (Zeng and Terada, 2001) and antioxidants (e.g., Vitamin E, alpha-tocopherol, glutathione, taurine, cysteine, butylated hydroxytoluene, superoxide dismutase and catalase) in the semen extenders does improve the freezing ability of spermatozoa of many species such as stallion (Aurich et al., 1997; Ball et al., 2001), bull (Beconi et al., 1993; Bilodeau et al., 2001), ram (Uysal and Bucak, 2007; Bucak et al., 2007), avian (Donoghue and Donoghue, 1997), boar (Cerolini et al., 2000) and some wildlife (Leibo and Songsasen, 2000). Studies have demonstrated that the supplement of alpha-tocopherol (Peña et al., 2003), butylated hydroxytoluene (Roca et al., 2004), superoxide dismutase and catalase (Roca et al.,

2005) in the semen extenders reduces the ROS formation and improve post-thawed sperm motility and viability of FT boar semen. In addition, it was found that the supplement of extended boar semen with 5 mM of cysteine improved the viability and functional status of the chilled boar spermatozoa (Funahashi and Sano, 2005).

2. History of cryopreservation

Nowadays, there are 2 techniques for cryopreservation in boar semen, traditional nitrogen method and controlled rate freezing method. Verheyen (1993) reported a significantly better post-thaw sperm outcome when computer controlled rate freezing was used compared to non-controlled rate freezing. In human, it has been reported that controlled rate freezer method provided significant superior post-thaw sperm motility, viability, and cryosurvival rate, compared with traditional nitrogen method (Petyim and Choavaratana, 2006). However, the study of Thalchil (1981) did not confirm the different outcome of these 2 methods. Besides breed-specific fertility, data from field trials found that the mean motility of frozen–thawed semen between Norwegian Landrace and Duroc boars was difference. In the different breeds of boars found the differences in membrane lipid composition, can neither explain the major differences in post-thaw survival and fertility between breeds (Waterhouse et al., 2006).

Major limitation of frozen-thawed semen (FT-boar semen) have been observed, i.e. low conception rate and low litter size after AI (Johnson et al., 2000; Buranaamnuay et al., 2006). The relatively low fertility of FT-boar is associated with many factors. It has been reported that reactive oxygen species (ROS) generation, induced by the cryopreservation process, can be responsible for mammalian sperm damage (Griveau and Le Lannou, 1997) ROS production has been associated with reduction of sperm motility and decreased capacity for sperm–oocyte fusion. Spermatozoa are sensitive to lipid peroxidation due to their high content of polyunsaturated fatty acids, and are unable of resynthesizing their membrane components, although this may not be the sole mechanism by which sperm function is impaired by ROS. Many studies have shown that the supplementation of antioxidants in extenders improved the qualities of both fresh boar semen (Bamba and Cran, 1992; Funahashi and Sano, 2005) and frozen boar semen (Breininger et al., 2005; Gadea et al., 2005; Pena et al., 2003; Roca et al., 2004; Roca et al., 2005). Earlier studies showed that the supplementation of some antioxidant such as, water soluble Vitamin E 200 μM to semen extenders for freezing boar spermatozoa reduced post-thaw ROS generation and improved sperm motility and viability (Pena et al., 2003). Funahashi and Sano (2005) reported that supplementation extended boar semen with glutathione or L-cysteine of 5 mM improved the viability and functional status of boar spermatozoa during liquid storage at 10 °C for at least 14 day. A recent report (Gadea et al., 2005) demonstrated that supplementation with 1 mM of reduced glutathione to freezing media resulted in a protective effect on sperm function.

3. Cryopreservation of animal spermatozoa

Cryopreservation of boar semen need to be developed for AI in the pig industry due to a number of reasons including preservation of a good genetic resource, increase genetic improvement, distribution of genetic lines across countries and reduce boar transportation

(Almlid and Hofmo, 1996; Johnson, 1998). The widespread exchange of genetic material between breeding populations with liquid stored semen is difficult because of the short life span of the spermatozoa (Wagner and Thibier, 2000; Johnson et al., 2000). The first FT boar spermatozoa have been reported since 1956 (Polge, 1956). Unfortunately, the FT spermatozoa has a very low fertilizing ability. In 1970, the first pregnancy was achieved with FT boar semen using a surgical insemination technique (Polge et al., 1970). In 1971, many studies have reported the pregnancies after intra-cervical insemination using FT boar semen in pig (Crabo et al., 1971; Pursel et al., 1971).

In general, there are many important factors in the process of FT boar semen that affect the post-thawed semen quality. For instance, the semen collection technique, equilibration time, type of semen extender, type and concentration of cryoprotectant, freezing package, freezing rate and thawing procedure (Johnson et al., 2000). Many types of freezing package have been used for frozen boar semen, such as medium straws, maxi straws (Bwanga et al., 1991; Berger and Fisherleitner, 1992), plastic bags (Bwanga et al., 1991) and FlatPack®/MiniFlatPack® (Eriksson and Rodriguez-Martinez, 2000). Most containers has been developed for suitable storage, transport, post thawed semen quality and practical insemination.

The freezing and thawing procedure have a significant impact on the survival rate of sperm after cryopreservation (Johnson et al., 2000). However, optimal freezing and thawing rates vary depending on the type and concentration of the cryoprotectant (Mazur et al., 1970; Fiser et al., 1993). Currently, the optimal rates for boar sperm freezing appear to be 30°C/min with 3% glycerol as cryoprotectant when freezing in 0.5 ml straws (Fiser et al., 1990) and 16°C/min with 3.3% glycerol in 5 ml straws (Pursel et al., 1985). For both these methods the optimal thawing rate is 1200 °C/min (Westerndorf et al., 1975; Fiser et al., 1993). Eriksson and Rodriguez-Martinez, (2000) found that the optimal freezing rate was 50 °C/min in 3% glycerol with a 900 °C/min thawing rate for flattened plastic bags (FlatPack®) container. A variety of cryoprotectants are used in the freezing extender of different in species. Glycerol, egg yolk and sodium dodecyl sulphate (SDS) (Equex STM or Orvus ES paste) is commonly used as cryoprotectants for the cryopreservation of boar semen (Westerndorf et al., 1975; Pursel et al., 1978; Holt, 2000b). The optimal concentration of glycerol was approximately 3 % in pig (Holt, 2000a). Egg yolk and SDS are non-permeable cryoprotectant used in freezing extender and provide protective effect to spermatozoa and improved post thawed sperm quality (Pursel et al., 1978). It has been suggested that SDS enhances the cryoprotective properties of the egg yolk to protect the sperm membrane from cryoinjuries (Buhr et al., 1996)

4. Boar semen cryopreservation methods

4.1 Semen collection

Three ejaculates from each boar are collected using gloved-hand method. During collection the semen is filtered through gauze and only sperm rich fractions are collected. Within 30 min after collection, semen volume, pH, sperm motility, concentration, percentage of live and dead sperm and morphology are determined. Only ejaculates with motility of ≥70% and ≥80 % morphologically normal are used for cryopreservation.

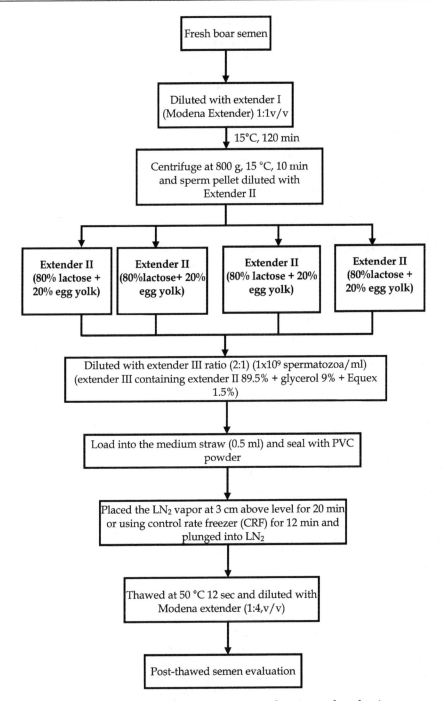

Fig. 1. Flow chart of the boar semen freezing processes, thawing and evaluation

4.2 Semen freezing and thawing procedures

Shortly after collection, the semen is diluted (1:1 v/v) with extender I (Modena™, Swine Genetics International, Ltd., Iowa, USA). The diluted semen is transferred to 50 ml centrifuge tubes, equilibrated at 15 °C for 120 min and centrifuged at 800x g for 10 min. The supernatant is discarded and the sperm pellet was re-suspended (about 1-2:1) with extender II (80 ml of 11% lactose solution and 20 ml egg yolk) to a concentration of $1.5x10^9$ spermatozoa/ml. The diluted semen is cooled to 5 °C for 90 min. Then, two parts of the semen are mixed with one part of extender III (89.5% of extender II with 9% glycerol and 1.5% Equex-STM®). The final concentration of semen is approximately $1.0x10^9$ spermatozoa/ml and contained 3% glycerol (modified after modified after Westerndorf et al., 1975 and Gadea et al., 2004). The processed semen is loaded into 0.5 ml straws (Bio-Vet, Z.I. Le Berdoulet, France). The straws are sealed with PVC powder before being placed in contact with nitrogen vapour about 3 cm above the liquid nitrogen level for 20 minutes in an expandable polystyrene box. Then the straws are plunged into liquid nitrogen (-196 °C) for storage. Thawing is achieved by immersing the straws in water at 50°C for 12 sec (Selles et al., 2003). Immediately after thawing, the semen is diluted (1:4) with a Modena™ extender. Post-thawed sperm qualities are evaluated after incubation in a 37°C water-bath for 15 min.

4.3 Semen extender

After incubation in extender I, the semen is divided into 4 groups according to the composition of extender II. Group I, extender II containing 80 ml of 11% lactose solution and 20 ml egg yolk. Group II, extender II is supplemented with 0.29 g of fish oil (Fish oil 1000; Blackmores LTD, New Southwell, Australia; containing DHA 120 mg/g fish oil) per gram of egg yolk. Normal egg yolk contains approximately 3.15 mg DHA per gram of egg yolk, as was analyzed at the Institute of Nutrition, Mahidol University (AOAC, 2007). Group III is supplemented with a combination of fish oil 0.29 g and L-cysteine 5 mM (Fluka Chemie GmbH, Sigma-Aldrich, Switzerland). Group IV is supplemented with a combination of fish oil 0.29 g and L-cysteine 10 mM.

5. Harmful effect of cryopreservation to spermatozoa

The use of FT boar semen under field conditions results in low conception rate and reduced number of total piglets born per litter (Eriksson et al, 2002). These problems occur because of the poor post-thawed semen quality and low survival rate of boar spermatozoa after cryopreservation (Hammerstedt et al., 1990; Curry et al., 2000). The detrimental effects of cooling, freezing and thawing caused subsequently impaired the membrane integrity, structure and function of the spermatozoa, eventually fertilizing ability (Hammerstedt et al., 1990; Guthrie and Welch, 2005).

It is well documented that the boar spermatozoa are highly susceptible to temperatures below 15 °C. The viability of the spermatozoa are dramatically reduced within a few hours after expose to cooling below 15 °C, so call 'cold shock' (Gilmore et al., 1996). Cold shock caused the damage of plasma membranes and alterations in the metabolism of the spermatozoa. This caused by changes in the arrangement of plasma membrane compositions especially phospholipids (reviewed by Medeiros et al., 2002).The sperm damage cause by cold shock is characterized by an irreversible loss of motility and the loss of sperm permeability. Boar spermatozoa seem to acquire a cold shock resistance when the semen is held at room temperature in seminal plasma for 1-5 hours (Pursel et al., 1972). It was found that viability and fertilizing ability of the boar spermatozoa was significantly improved when the fresh semen was held at 15 °C for over 3 hours before cryopresenvation (Almlid and Johnson, 1988; Eriksson et al., 2001)

During the process of freezing, the decrease of temperature from -15 °C to -60 °C causes sperm damage (Mazur, 1985). This causes by the intracellular ice formation and cellular dehydration (osmotic stress). The subsequent physical events depend on the cooling rates. Intracellur ice formation occur during a rapid cooling when intracellular water does not leave the cell to maintain equilibration. If cooling is slow, the spermatozoa will lose water rapidly avoid to intracellular ice formation. However, if spermatozoa are cooled too slowly, they will expose to high concentration of solutes which caused intracellular water to diffuse out of the cell, dehydration both the cell and plasma membrane (also known as solution effects) (Mazur, 1970; Parks and Graham, 1992). Gilmore et al. (1996) demonstrated that the boar spermatozoa are sensitive to osmotic stress. Similar findings were also found in dog (Songsasen et al., 2002), cat (Pukazhenthi et al., 2000), ram (Curry and Watson, 1994), stallion (Ball and Vo, 2001) and bull (Liu and Foote, 1998).

Cellular damage due to intracellular ice formation and dehydration, oxidative stress is another important cause of sperm damage leading to abnormal sperm structure and function and subfertility. Currently, several studies have been reported oxidative stress affect the damage of sperm membrane, proteins and DNA in human (Agarwal et al., 2003), stallion (Baumber et al., 2000; 2003), bull (Bilodeau et al., 2001) and boar (Roca et al., 2004; 2005)(Fig. 2).

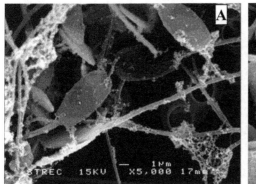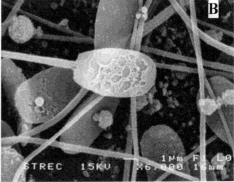

Fig. 2. Scanning electron microscopic (SEM) picture of fresh boar semen with normal plasma membrane (A) as compare with SEM picture of frozen boar semen with plasma membrane damage (B).

6. Capacitation like changes of the frozen-thawed boar semen and its influence on *in vivo* fertility

After ejaculation, the spermatozoa are not able to fertilize the oocyte. They undergo the activation process that called ' capacitation' and acrosome reaction which spermatozoa will able to reach the ampulla of the oviduct, penetrate the cumulus oophorus, bind to the zona pellucida, activate the acrosome reaction and eventually fertilize the oocyte (Yanagimachi, 1994). Normally, capacitation *in vivo* occurs in the female reproductive tract, but capacitation can also be induced in vitro by the incubation in capacitating media which the most of media contain the bicarbonate, calcium and serum albumin (Yanagimachi, 1994). Furthermore, It has been demonstrated that the cooling and freezing process can also induced the capacitation like change which affect to the low fertilizing capacity of spermatozoa in boar and other mammalian species (Maxwell and Johnson, 1997; Green and Watson, 2001; Barrios et al., 2000). During cooling, freezing and rewarming process, it is hypothesized that change in low temperature cause the modification and destabilization of the lipid content in the sperm plasma membrane, reducing the selective permeability resulted in the cholesterol efflux and intracellular calcium uptake leading to the capacitation like change (Green and Watson, 2001; Tardif et al., 2001). To improve the the FT spermatozoa, there are some studies about the addition of cholesterol-loaded cyclodextrins increased the cryosurvival of boar, ram and bovine spermatozoa because cyclodextrins used to deliver cholesterol to the sperm plasma membrane which against cold shock (Purdy and Graham, 2004; Bailey et al., 2008; Mocé et al., 2009). In addition, the supplement of antioxidants such as vitamin E or alpha-tocopherol decreased the capacitation like change of cryopreserved boar spermatozoa (Satorre et al, 2007). Furthermore, the addition of seminal plasma to boar spermatozoa has been shown to reduce the capacitated spermatozoa in chilled and FT boar semen (Kaneto et al., 2002; Suzuki et al., 2002; Vadnais et al., 2005a,b; Okazaki et al., 2009, Garcia et al., 2009).

7. Laboratory method for semen quality assessment

7.1 Sperm concentration and progressive motility

Sperm concentration will be assessed by direct cell count using a Bürker haemocytometer (Boeco, Humburg, Germany) (Beardon and Fuquay, 1997). The visual progressive motility of both fresh and FT sperm is evaluated at 38°C under a phase contrast microscope at 200x and 400x magnification. The motility is assessed by the same person throughout the experiment.

7.2 Computer-assisted sperm analysis (CASA)

The motility patterns of diluted FT semen are assessed using the CASA system (Halminton Thorne Biosciences IVOS, Version 12 TOX IVOS, Beverly, USA). Each FT thawed semen samples is diluted with pre-warmed Modena extender (37°C) to obtain a final concentration of 50×10^6 spermatozoa/ml. A 5 µl of diluted semen is pipetted into the chamber and allowed the 1 min before analysis fore sample distribution and pre-warming (Iguer-Ouada and Verstegen, 2001). After the first assessment (T0), the diluted semen is evaluated after incubation at 37 °C for 30 min (T30) and 60 min (T60). The camera will recognize the position of the sperm heads in successive frames. Spermatozoa heads are marked with a different color to enable the observer and the analyzer to differentiate between the different motility patterns. Each semen sample is measured twice, 3 fields are evaluated and counted at least 1000 cells per analysis. Motility patterns including (1) Curvilinear velocity (VCL, µm/s), the average velocity measured in the progression line along the whole track of cell path; (2) Average pathway velocity (VAP, μm/s) , the average velocity of the smoothed cell path; (3) Straight line velocity (VSL), the average velocity measured in a straight line from the beginning to the end of the track (μm/s); (4) The amplitude of the lateral head displacement (ALH), the mean width of the head oscillation as the sperm cells swim (μm); (5) The beat cross-frequency (BCF, Hz), frequency of the sperm head crossing the average path in either direction; (6) The straightness (STR, %) = average value of the ratio VSL/VAP; (7) The Linearity (LIN, %) = average value of the ratio VCL/VAP.

7.3 Sperm viability

The percentages of sperm viability will be determined by 2 methods. The first one is eosin-nigrosin staining (Dott and Foster, 1972). The semen sample (50 μl) are mixed well with a drop of eosin-nigrosin dyes (Fluka Chemie GmbH, Sigma-Aldrich, Switzerland), and the mixture (10 μl) is smeared and dried on a glass slide. Evaluation is undertaken by counting 200 spermatozoa with 1000x magnification. Spermatozoa with an unstained head are regarded as live spermatozoa. The second method is evaluated by SYBR-14/Ethidiumhomodimer-1 (EthD-1) (Fertilight®, Sperm Viability Kit, Molecular Probes Europe, Leiden, The Netherlands). This technique is modified after Axnér et al. 2004 and Garner and Johnson, 1995). Ten μl of diluted semen are mixed with 2.7 μl of the user solution of SYBR-14 and 10 μl of EthD-1. The user solution is SYBR-14 diluted (1:100) in dimethyl sulfoxide (DMSO), fractionated and frozen in eppendorfs. After incubation at 37 °C for 20 min, two hundred spermatozoa will be assessed (x1000) under fluorescent microscope. The nuclei of the spermatozoa with an intact plasma membrane are stained green with SYBR-14, while those with damaged membranes stained red with EthD-1.

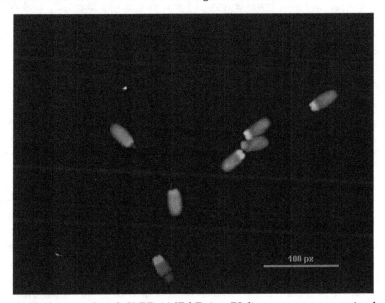

Fig. 3. Spermatozoa stained with SYBR-14/EthD-1 or PI: live spermatozoa stained green with SYBR-14 while dead spermatozoa stained red with EthD-1 or PI.

Spermatozoa are classified into three types; live spermatozoa stained green with SYBR-14, dead spermatozoa stained red with EthD-1 and moribund spermatozoa stained both green and red (Axnér et al. 2004; Garner and Johnson, 1995).The results are expressed as the percentage of live spermatozoa with intact plasma membranes.

7.4 Acrosome integrity

Acrosome integrity will be assessed using fluorescein isothiocyanate–labeled peanut (*Arachis hypogaea*) agglutinin (FITC-PNA) staining. Ten µl of the diluted semen is mixed with 10 µl of Ethidiumhomodimer-1 and incubated at 37 °C for 15 min. Five µl of the mixture is smeared on a glass slide and fixed with 95 % ethanol for 30 second. Fifty µl Fit C-PNA (dilute Fit C-PNA with PBS 1:10 v/v) is spread over the slide and incubated in a moist chamber at 4 °C for 30 min. After incubation, it is rinsed with cold PBS and air dried. Two hundred spermatozoa are assessed under fluorescent microscope at 1000x magnification and classified as intact acrosome, damaged acrosome and missing acrosome (Cheng et al., 1996; Axner et al., 2004). The results are scored as the percentage of intact acrosome spermatozoa.

7.5 The functional integrity of the sperm plasma membrane

The functional integrity of the sperm plasma membrane will be assessed using a short hypo-osmotic swelling test (sHOST) (Perez-Llano et al., 2001). Spermatozoa are incubated, at 38 °C for 30 min, with 75 mOsm/kg a hypo-osmotic solution that consist of 0.368 % (w/v) Na-citrate and 0.675 % (w/v) fructose (Merck, Germany) in distilled water. Following this incubation time, 200 µl of the semen-hypo-osmotic solution is fixed in 1000 µl of a hypo-osmotic solution plus 5 % formaldehyde (Merck, Germany), for later evaluation. Two hundred spermatozoa are assessed under a phase contrast microscope at 400x magnification. The coiled tail (sHOST positive) spermatozoa found following incubation are functional intact plasma membrane.

7.6 DNA damage

DNA damage can be evaluated by Acridine orange (AO) staining or Halomax staining method (Fig. 4). The technique is modified after Thuwanut et al. (2008). Briefly, two smears from each sample were prepared on glass slide and air-dried. Each smear is fixed overnight in Carnoy's solution, freshly prepared with methanol and glacial acetic acid (3:1 v/v). The slide is removed from the fixative solution, air-dried, and then stained with 1% (100 mg/ml) AO (Sigma) in distilled water for 10 min. The AO staining solution is prepared by adding 10 ml of 1% AO in distilled water to 40 mL of 0.1 M citric acid (Merck, Darmstadt, Germany) and 2.5 ml of 0.3 M Na2HPO4.7H2O (Merck, Darmstadt, Germany) pH 2.5. The AO staining solution will be prepared daily and stored in the dark at room temperature until use. After staining, the slide is gently washed by distilled water and covered with the cover slip. One thousand spermatozoa are evaluated under the fluorescence microscope. The heads of the sperm cells with normal DNA (double-stranded) have green fluorescence, while those with damaged or single stranded DNA showed orange or red fluorescence. The results are expressed as the proportion of the damage/single stranded DNA per 1,000 counted spermatozoa.

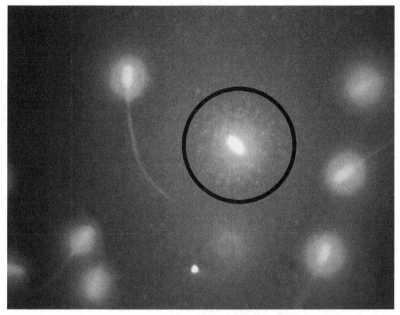

Fig. 4. Sperm DNA damage (black circle) stained with a commercial kit (Halomax staining)

7.7 Chlortetracycline (CTC) assay

The CTC assay is slightly modified from as described previously (Harayama et al., 2000). The CTC staining solution containing 750 μm CTC, 5 mM DL-cysteine, 130 mM NaCl and 20mM Tris (hydroxymethyl aminomethane) (pH 7.8) is prepared immediately before use. This solution is protected from light until analysis. Briefly, 50 μl of sperm suspension will be mixed with 50 μl of CTC staining solution for 30 sec, followed by the addition of 10μl 12.5% paraformaldehyde in 0.5 M Tris-HCl (pH 7.4) as a fixative. Then, 10 μl of the sperm suspension is mixed well with equal volume of antifade solution (0.22 M of 1,4-diazabicyclo[2,2,2]octane ;DABCO) in glycerol:PBS (9:1) on the microscopic slide and gently compressed with coverslip. Two slides are prepared from each sample and stored in the dark at at 4 °C until evaluation. Two hundred spermatozoa per slide will be under a Nikon fluorescence microscope at 400x under blue-violet illumination (excitation at 400–440 nm and emission at 470 nm). The spermatozoa are classified in to three staining patterns described by Fraser et al. (1995). F-pattern, fluorescence over the whole region of the sperm head are considered to be "non-capacitated spermatozoa". B-pattern, fluorescence in the acrosomal region except post-acrosomal region are considered to be "capacitated spermatozoa". AR-pattern, low or no fluorescence over the whole head except thin bright ban in the equatorial segment are considered to be "acrosome-reacted spermatozoa".

7.8 Annexin-V/PI assay

Apoptosis will be evaluated by apoptosis detection kit ApopNexin™ (Chemicon Int., USA) using a fluorescent microscope. This assay will detect the phosphatidylserine translocation from inner to outer leaflet of cell plasma membrane which is the hallmark of apoptosis during the degradation phase. Following manufacturer instructions, sperm cells are washed twice with PBS (pH 7.4) by centrifugation at 400 g for 5 min. Sperm pellet are resuspended with HEPES buffer (10mM HEPES/NaOH, pH7.4, 150mM NaCl, 5mM KCl, 1mM MgCl2, 1.8mM CaCl2, 2x106 sperm/ml). One hundred µl of sperm suspension are mixed well with 5 µl annexinV-FITC conjugate and 3 µl of Propidium iodide (PI;20µg/ml) and incubated for 15 min at room temperature in the dark. Two hundred spermatozoa are assessed under fluorescent microscope at 400x magnification. The apoptotic sperm cells will fluorescence green while necrotic sperm cells fluorescence red. Alternatively, flow cytometry analysis can also be used instead of fluorescent microscope (Fig. 5).

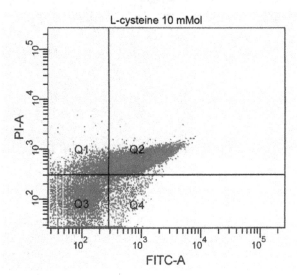

Fig. 5. FCS/SCC two-dimensional histogram, flow cytometry analysis of frozen-thawed boar spermatozoa (supplemented with L-cysteine) stained with Annexin-V/PI: Q3 represent the viable spermatozoa with intact plasma membrane, while Q2 represent dead spermatozoa with non-intact plasma membrane

8. Lipid composition of sperm plasma membrane

The lipid compositions of the plasma membrane of the mammalian spermatozoa are markedly different from those of somatic cells. In general, the sperm plasma membrane contains approximately 70% phospholipids, 25% neutral lipids, and 5 % glycolipids (Flesch and Gadella, 2000). All lipid components located in the sperm membranes responsible for the fluidity of membrane lipid bilayers, regulation of sperm maturation, spermatogenesis, capacitation, acrosome reaction and membrane fusion (Parks and Hammerstedt, 1985; Martinez and Morros, 1996; Sanocka and Kurpisz, 2004).

Sperm plasma membrane are made up of a phospholipids bilayer, with the major phospholipids were choline phosphoglycerides (CP), ethanolamine phosphoglycerides (EP) and sphingomyelin which their proportions differed between species. These phospholipids contain a high proportion of long chain, polyunsaturated docosapentanoyl (22:5) and docosahexanoyl (22:6) groups which both lipids represent approximately 50 to 60 % of total phospholipids in boar and bull spermatozoa (Pursel and Graham, 1967; Johnson et al., 1969; Parks and Lynch, 1992). Cholesterol was the major sterol in sperm lipids of all species. Cholesterol to phospholipid molar ratios were 0.26, 0.30, 0.36, and 0.45 for sperm plasma membrane of the boar, rooster, stallion, and bull, respectively (Parks and Lynch, 1992).Glycolipids represented less than 10% of total polar lipids for all species.

The susceptibility of spermatozoa to cold shock differ among species because of the differences of lipid composition of the sperm plasma membrane among species (Flesch and Gadella, 2000).The resistance of the mammalian spermatozoa to cold shock was high in species in which the cholesterol to phospholipids molar ratio and the phospholipids saturation is high (Darin-Bennett and White, 1977). The avian spermatozoa have a high level of cold shock resistant and have a higher level of saturated phospholipids compared to mammalian sperm (Parks and Lynch, 1992). The plasma membrane of the boar spermatozoa is characterized by a high protein, low cholesterol and high proportion of EP compared to other species (Parks and Lynch, 1992; Nikolopoulou et al., 1985). In contrast, the protein content and EP proportion of rooster sperm plasma membrane is low while the cholesterol content is intermediate (Parks and Lynch, 1992).

As mentioned above the sperm plasma membrane has a very high amounts of polyunsaturated fatty acids (PUFAs) especially docosapentaenoic acid (DPA) and docosahexaenoic acid (DHA) (Johnson et al., 1969; Parks and Lynch, 1992). It has been suggested that the proportion of unsaturated fatty acid influence the properties of sperm plasma membrane (Miller et al., 2005). High levels of long chain PUFAs, DPA and DHA, are associated with an increased membrane fluidity (Quinn, 1985). During cryopreservation, the fluidity of the plasma membrane from boar spermatozoa is significantly decreased when compared to fresh spermatozoa which tend to restrict the post-thawed sperm quality (Buhr et al., 1994). In human, sperm with a high level of membrane fluidity had a higher post-thawed motility compared to sperm with a low level of membrane fluidity after cryopreservation (Giraud et al., 2000).

9. Docosahexaenoic acid (DHA)

Docosahexaenoic acid (commonly known as DHA; 22:6 (n-3)) is an omega-3 essential polyunsaturated fatty acid. DHA is most often found in cold water fatty fish (salmon fish, tuna fish) and in fish oil supplements, along with eicosapentaenoic acid (EPA). DHA is the main fatty acid composition of the spermatozoa as well as the brain and the retina (Neuringer et al., 1988). For the sperm plasma membrane, DHA play a major role in regulating membrane fluidity in sperm and in the regulation of spermatogenesis (Haidl and Opper, 1997; Ollero et al., 2000). DHA content is significantly higher in immature spermatozoa than mature spermatozoa.

Studies have demonstrated that the supplement of PUFAs in the feed of the boar improve the quality of the boar spermatozoa (Paulenz et al., 1999; Rooke et al., 2001; Strezezek et al., 2004 ;Maldjian et al., 2005). In addition, Rooke et al. (2001) found that tuna oil supplemented in the boar diet increase viability, progressive motility and normal morphology. The supplementation of PUFAs also enhanced the survival rate of post-thawed boar spermatozoa (Strezezek et al., 2004). DHA improved the reproductive performance of the male turkey (Blesbois et al., 2004). Maldjian et al. (2005) found that the use of DHA-enriched hen egg yolk for the semen extender and the supplement of 3% fish oil in the boar feed increased the DHA content of the boar spermatozoa post-thawed. However, the authors could not demonstrate the improvement of the quality of post-thawed spermatozoa.

10. Oxidative stress and sperm function

Oxidative stress is a condition associated with an increasing rate of cellular damage, induced by oxygen and oxygen-derived oxidants, commonly known as ROS (Sikka et al., 1995). ROS are highly reactive oxidizing agents belonging to the class of free-radicals, which contains one or more unpaired electrons. Normally ROS included superoxide anion (O_2^-), hydrogen peroxide (H_2O_2), peroxyl radical (ROO-) and the very reactive hydroxyl radicals (OH-). The nitrogen-derived free radical nitric oxide (NO) and peroxynitrite anion (ONOO-) play an important role to the fertilization. Two main resources of ROS in semen include leukocytes and immature or defective spermatozoa (Aitken et al., 1992; Silva, 2006).

It is known longtime ago that ROS harm the spermatozoa (Macleod, 1943). Nowadays, studies have shown that the cryopreservation of spermatozoa induces the ROS formation and led to lipid peroxidation as well as DNA oxidation. These events attribute to the decrease of sperm function and infertility (Aitken et al., 1991; Alvarez and Storey, 1992; Agarwal, 2003). Nevertheless, spermatozoa normally produce a small amounts of ROS, needed for capacitation and acrosome reaction (Agarwal et al., 2005; De Lamirande and Gagnon, 1993).

Oxidative stress is the result of an imbalance between ROS generation and scavenging activities (Sikka et al., 1995; Sharma and Agarwal, 1996). Spermatozoa are sensitive to oxidative stress because of low concentrations of scavenging enzymes in the cytoplasm (de Lamirande and Gagnon, 1995; Saleh and Agarwal, 2002) and the plasma membranes contain high amounts of PUFAs (Alvarez and Storey, 1995). ROS act as triggers a chain of reaction. Lipid peroxidation (LPO) (De Lamirande and Gagnon, 1992; Sikka et al., 1995). LPO of sperm plasma membrane is the key mechanism of ROS-induced sperm damage (Alvarez et al., 1987)..??.

LPO of sperm membranes is an autocatalytic self-reaction composed of 3 steps. Firstly, initiation step, this is the abstraction of a hydrogen atom from an unsaturated fatty acid. Secondly, propagation step, this is the formation of alkyl radical which followed by its rapid reaction with oxygen to form a peroxyl radical is capable of abstracting a hydrogen atom from an unsaturated fatty acid with the concomitant formation of a lipid radical and lipid peroxide such as hydrogen peroxide(H_2O_2). Since the peroxyl and alkyl radicals are regenerated, the cycle of propagation could continue indefinitely. Finally, the termination step, the substrates is consumed or stopped by the radical-radical reaction which produce a non-radical species (Sanocka and Kurpisz, 2004). LPO has been reported to affect the sperm dysfunction associated with decreased membrane fluidity, loss of membrane integrity and function of spermatozoa (Sanocka and Kurpisz, 2004). Furthermore, LPO also damage DNA and proteins resulted in an increased the susceptibility to be attacked by the macrophage (Aitken et al., 1994).

10.1 Effect of antioxidants on oxidative stress and sperm function

Antioxidants are compounds that suppress the formation of ROS and protect spermatozoa against ROS (Sikka, 1995). Studies have demonstrated that seminal plasma contains a number of enzymatic antioxidants such as superoxide dismutase (SOD; Alvarez et al., 1987), glutathione peroxidase/glutathione reductase (GPX/GRD) and catalase. These antioxidants protect the spermatozoa against LPO (Lenzi et al., 1996; Sikka et al., 1996; Saleh and Agarwal, 2002). SOD spontaneously dismutates (O_2^-) anion to form O_2 and H_2O_2. Catalase converts H_2O_2 to O_2 and H_2O. In addition, glutathione peroxidase, a selenium-containing antioxidant enzyme with glutathione, is an electron donor removes peroxyl (ROO^-) radicals from various peroxides including H_2O_2 (Sikka et al., 1996). In addition, seminal plasma contains a variety of non-enzymatic antioxidants such as ascorbic acid (vitamin C), alpha-tocopherol (vitamin E), and reduced glutathione (Lenzi et al., 1994;Saleh and Agarwal, 2002; Silva, 2006).

Vitamin C is a major chain–breaking antioxidant present in the extracellular fluid (Saleh and Agarwal, 2002). It neutralized hydroxyl, superoxide and hydrogen peroxide radicals and prevent sperm agglutination (Agarwal et al., 2004). Vitamin E is a chain-breaking antioxidant in the cell membrane, inhibits LPO by scavenging peroxyl and alkoxyl radicals. Glutathione is the most abundant antioxidant , plays a role in protecting lipids, proteins and nucleic acids against oxidative stress.

Studies have shown that the supplementation of antioxidants in extenders both chilled and frozen-thawed semen such as alpha-tocopherol, butylated hydroxytoluene, superoxide dismutase and catalase, cysteine or glutathione have been reported to improve the semen quality in boar (Pursel, 1979; Bamba and Cran, 1992; Brezezinska-Slebodzinska E, 1995; Cerolini et al., 2000; Penã et al., 2003; Gadea et al., 2004, Roca et al., 2004, 2005; Funahashi and Sano, 2005; Breininger et al., 2005; Satorre et al., 2007), bull (Bilodeau et al., 2001), turkey (Donoghue and Donoghue, 1997), stallion (Aurich et al., 1997; Ball et al., 2001) and ram (Uysal and Bucak, 2007).

10.2 Effect of L-Cysteine on frozen boar semen

L-cysteine, an amino acid containing a sulphydryl group, is a precursor of intracellular glutathione biosynthesis. L-cysteine plays a role in the intracellular protective mechanism against oxidative stress, membrane stabiliser and capacitation inhibitor (Johnson et al., 2000). Glutathione is the most common non-thiol protein in mammalian cells which protects plasma membrane from LPO, scavenges superoxide and minimized O_2^- formation. It has been demonstrated that the supplementation of L-cysteine in the semen extender prevents the loss of sperm motility by minimizing hydrogen peroxide of FT semen in the bull (Bilodeau et al., 2001). Funahashi and Sano (2005) found that the supplement of L-Cysteine for 5 mM improved the viability and functional status of the boar spermatozoa during chilled storage.

During the past few years, many studies have been carried out by supplementation of various antioxidants (e.g. Vitamin E, Glutathione, Taurine) in the freezing extenders of frozen boar semen in order to minimize the detrimental effect of ROS which occurred during the freezing process (Pena et al., 2003; Roca et al., 2004; Breininger et al. 2005; Gadea et al., 2005). Funahashi and Sano (2005) demonstrated that supplement of L-cysteine (5 mM) could improve the viability and progressive motility in fresh boar semen, and also the same case found in frozen bovine semen (Bilodeau et al., 2001). This L-cysteine is also improve survival time of semen and sperm chromatin structure in fresh chilled boar semen at 15ºc

(Szczesniak-Fabianczyk et al. 2003). In frozen dog semen, Micheal et al. (2007) reported that supplement of L-cysteine resulted in increased viability and rapid steady forward movement (RSF movement). Recently, Kaeoket et al. (2008b) also found that addition of 5 mM L-cysteine (the same concentration used for fresh boar semen preservation by Funahashi and Sano, 2005) has a tendency (not significant difference) to improve post-thawed semen quality when compare with the addition of glutathione and water-soluble vitamin E.

10.3 Effect of L-Cysteine x DHA on frozen boar semen

The characteristics of the sperm cryoinjury included the destabilization of lipid bilayer of the sperm plasma membrane, change in permeability of plasma membrane and a reduction of the viability of sperm. It is well documented that the boar sperm are highly susceptible to temperature below 15 °C mainly due to a relatively poor lipid composition and structure of plasma membrane compared to other domestic species. The boar sperm plasma membrane consisted of a high levels of polyunsaturated fatty acids (PUFAs) especially docosapentaenoic acid (DPA) and docosahexaenoic acid (DHA) and low level of cholesterol: phospholipids ratio. It has been shown that the level of PUFA content play an important role in the sperm membrane fluidity and cause sperm susceptible to lipid peroxidation (LPO). During frozen-thawed (FT) process, sperm are attacked by reactive oxygen species (ROS) owing to LPO and leading to significantly decrease in the PUFAs content of their plasma membrane. ROS were mainly produced by the defective or dead spermatozoa and result in a reduction of sperm motility, sperm viability and eventually fertilizing ability. In order to minimize the sperm cryoinjury, the supplement of antioxidant compounds and some fatty acid to the semen extender have been reported to minimize ROS formation and enhance the plasma membrane function in many species. L-cysteine, a precursor of intracellular glutathione, plays an important role in the protecting sperm from oxidative stress and act as capacitation inhibitor. Earlier studies have demonstrated that L-cysteine supplement in the semen extender improve the motility of FT bull semen, prolonged sperm survival time and reduced chromatin damage in FT boar sperm. In addition, the use of n-3-enriched hen egg yolk in the semen extender increased the proportion of DHA content in the boar sperm. Our previous study found that the addition of L-cysteine directly into lactose egg yolk (DHA-enriched) base extender significantly improves the sperm motility and intact acrosome of FT boar sperm. In addition, Kaeoket et al. (2010) found that the supplement of DHA (fish oil) improves the sperm motility, viability and acrosome integrity of the FT boar sperm.

10.4 Effect of seminal plasma on post-thawing semen quality and reproductive performance after artificial insemination

Seminal plasma is the liquid constituent of an ejaculate, comprising a combination of fluids secreted by the male accessory glands (i.e., mainly from the seminal vesicle in boars) during an ejaculation. There is evidence that seminal plasma is able to arrest or reverse cryoinjury and perhaps extend the longevity of the sperm by inhibiting or reversing capacitation and acrosome reactions, and also by its antioxidant activity (Brzezińska-Ślebodzińska et al. 1995; Strzezek et al. 1999; Suzuki et al. 2002; Vadnais and Roberts 2007; Bailey et al. 2008). During the cryopreservation process of boar semen, seminal plasma is normally not required, and discarded by the centrifugation at the beginning of semen preparation, which may result in a lack of a significant contribution (i.e., antioxidant property, inhibiting or reversing capacitation, inhibiting and acrosome reaction and binding ability of its protein to the sperm plasma membrane) of seminal plasma in protecting sperm from cryoinjury. It has been demonstrated in rams that seminal plasma protein is able to revert the cold shock damage on the sperm membrane (Barrios et al. 2000). This effect has also been reported in boars, in that holding boar spermatozoa in its seminal plasma before cooling and freezing defends against the cold shock (Pursel et al. 1973). In addition, it has been shown that adding seminal plasma to the post-thawing solution increased the percentage of stallion sperm motility (Alghamdi et al. 2005). In dogs, it has been shown that frozen semen diluted with their prostatic fluid yielded a higher post-thawing motility (Rota et al. 2007). As a result, it seems likely that seminal plasma or prostatic fluid constituents have a positive effect on post-thawing sperm motility. In addition, semen with a high percentage of progressive motility illustrates their plasma membrane integrity and superb metabolism (Johnson et al., 2000) and also reflects their ability for fertilization (Vyt et al. 2004; Estienne et al. 2007). It can be hypothesized that the presence of supernatant (semen plasma plus semen extender) during the thawing process may improve frozen-thawed boar sperm motility.

Seminal plasma is the liquid constituent of an ejaculate, comprising a combination of fluids secreted by the male accessory glands (i.e. seminal vesicle) during an ejaculation. It is evidence that seminal plasma able to arrest or reverse cryoinjury and perhaps extend the longevity of the sperm by inhibiting or reversing capacitation and acrosome reaction (Suzuki et al., 2002). Besides, seminal plasma appears to play an important role in the female reproductive tract after insemination, e.g. attenuate the post insemination inflammatory response in the uterus of the sow which may influence the chances of conception (Rozeboom et al.,1999) and its component such as hormone estrogen may also resulted in a release of prostaglandins from the pigs endometrium to the utero-ovarian veins and lymphatic vessels which in turn decrease duration time from standing oestrus to ovulation in gilts (Clause et al.,1987,1990; Weitze et al.,1990), Therefore, it seems likely that seminal plasma constituents (both the oestrogen and the protein fraction) have an effect on ovulation time in sows (Waberski et al.,1995; Kaeoket and Tummaruk, 2002b).

Generally, limitations to achieve a high reproductive performance in swine arise from a failure of sows to express estrus, failure to accurately determine onset of estrus for artificial inseminations and failure to determine the ovulation time after standing oestrus. The best predictor for time of ovulation is frequent detection of oestrus, because time of ovulation occurs approximately 38 to 48 h after onset of estrus (Anderson et al., 1993; Weitze et al., 1994; Soede et al., 1995). The effects of the timing of insemination relative to ovulation on fertilization rate has been study by Soede et al.(1995).Targeting insemination within 24 h before ovulation seems optimal to achieve a high farrowing rate and large little size (Nissen et al., 1997). Insemination between 0 and 24 h before ovulation results in high fertilization rates and consequently, a low number of re-breeders and a slightly higher litter size (Kemp et al., 1996; Kaeoket et al., 2002a, 2005).

10.5 Effect of long term versus short term extenders as freezing extender I on quality of frozen boar semen

In pig industry, the boar semen used for artificial insemination is extended with semen extender and kept in cold storage at 18-20°C for few days before artificial insemination. It has been recently reported that using of long term extenders (i.e. Androstar®Plus,

Modena™, Vitasem LD) to preserve fresh semen for 7 days yield a superior fresh boar semen qualities compare with those using of short term extender (Kaeoket et al., 2010d). In addition, the different in extended fresh semen qualities were also found depending on each type of long term extender used. This indicated that some constituents in each long term extender may assist sperm to overcome cold shock during cold storage. Generally, the difference between the short term and long term extenders are the ingredients contained in the extenders. Long term extenders contain complex buffering agent (i.e. HEPES, Tris, TES and MOPs) and antioxidants (i.e., bovine serum albumin (BSA), beta-carotene, cysteine, taurine, vitamin E and ascorbic acid) (Alvarez and Storey, 1995; Gadea, 2003; Funahashi and Sano, 2005), which can maintain semen qualities during cold storage for a longer period than short term extender.

10.6 Effect of differents sugars in LEY freezing extender on frozen boar semen quality

During cryopreservation, both physical and chemical factors including the rapid change in temperature or thermal stress, the intracellular ice formation, oxidative stress and osmotic stress led to the sperm plasma membrane damage (Meideros et al., 2000). Generally, the freezing extender consists of cryoprotectant, sugars, buffer, and some antibiotics (Johnson et al., 2000). Glycerol is the most common permeable cryoprotectant used for cryopreservation of boar semen (Holt, 2000). Egg yolk is a common non-permeable cryoprotectant. Different types of sugars, such as, trehalose, lactose, fructose, have been used in the freezing extender of boar semen (Purdy et al., 2006). Sugar is not only a source of energy but also protects the spermatozoa from dehydration and intracellular ice formation during the cryopreservation process (Watson, 2000). In general, lactose is the most common sugar used for the cryopreservation of boar semen (Johnson, 1985; Buranaamnuay et al., 2009; Chanapiwat et al., 2009; Chanapiwat et al., 2010; Kaeoket et al., 2010a; Kaeoket et al., 2010b; Kasettrut and Kaeoket, 2010). The effect of either type or concentration of sugar supplement in the freezing extender on the post-thawed semen qualities has been reported in dog (Yildiz et al., 2000; Yamashiro et al., 2007), ram (Aisen et al., 2002), bovine (Woelders, et al., 1997; Hu et al., 2010) and boar (Roca et al., 2008; Gutiérrez-Pérez et al., 2009; Malo et al., 2010; Mercado et al., 2010). For instance, the supplement of 55 mM glucose improved the motility pattern of the FT boar spermatozoa compared to 0 and 180 mM (Roca et al., 2008). Hu et al. (2009) found that the addition of 100 mM trehalose in the extender improved post-thawed boar sperm motility, viability and acrosome integrity compared to 0, 25, 50 and 200 mM. In addition, Malo et al. (2010) found that the trehalose-based freezing extender enhances the sperm survival rate and the fertilization rate by *in vitro* fertilization (IVF) compared to lactose and glucose based freezing extender. In order to improve the post-thawed boar sperm quality, it is important to investigate the influence of different sugars on FT boar sperm.

11. Phytosterol on frozen boar semen quality

During the past decade, several studies have focused on supplementation with a variety of antioxidants (e.g. vitamin E, vitamin C, L-cysteine, glutathione, taurine, pyruvate, SOD, catalase) in the freezing extenders of frozen boar semen with an attempt to minimize the detrimental effects of ROS, which occur during the freezing process.

Gamma-oryzanol, a phytosteryl ferulate mixture extracted from rice bran oil, has received a great deal of attention because of its significant various health-promoting functions such as antioxidant activity, inhibition of lipoperoxidation by its scavenging activity, reduction in LDL cholesterol and induction of HDL cholesterol, inhibition of platelet aggregation [20], its potential implications as a UV-A filter in sunscreen cosmetics, treatment of type 2 diabetes mellitus and allergic reactions. These data suggest that gamma-oryzanol, especially, with its antioxidant and scavenging activities can be useful as an antioxidant and lipid peroxidation inhibitor (i.e., membrane stabilizing) during cryopreservation. Rice bran oil is widely used in salad dressing and cooking oil in Asian countries including China, India, Japan and Thailand. At present, it is becoming to gain acceptance in Western countries as well. However, no scientific information is available on its antioxidant and scavenging activities in minimizing the detrimental effects of ROS during the cryopreservation of semen.

12. Artificial insemination with frozen boar semen

In pigs, it is well-documented that the optimal insemination time for fresh semen to maximize the good fertilisation rate is within 24 h before ovulation. It has been shown that the fertile life span of the pig oocyte is limited to between 8-12 h after ovulation. At suboptimal times for artificial insemination (AI) leads to inferior FR and litter sizes results. In addition, Kaeoket et al. (2002; 2005) demonstrated that when sows were inseminated after ovulation, fertilised oocytes and developed embryos were observed up to Day 11 but no embryos were found at Day 19. Subsequently, these sows returned to oestrus with a prolonged interval.

It has been demonstrated that the duration of oestrus is related to the WOI, i.e., sows with a short WOI (3-4 days) on average have a long oestrus duration, which is associated with a longer time from onset of oestrus to ovulation. On the contrary, sows with a WOI of 5-6 days or longer, have a shorter time from onset of oestrus to ovulation, and therefore should be inseminated (with fresh semen) sooner after the onset of oestrus to ensure that the first insemination occurs before ovulation. This recommendation is in accordance with the observation that the average timing of ovulation varies between 64 and to 72% of the duration of oestrus.

For frozen boar semen, it is predictable that insemination with frozen-thawed semen will result in lower PR, low FR and litter sizes. During the last decade, most of the experiments with fertility tests (field trial) of frozen-thawed boar semen have been carried out by using deep intrauterine insemination (DIUI, dose ranged from 150 million to 1 billion spermatozoa). Nevertheless, for fresh semen, an intrauterine insemination (doses ranging from 1-3 billion) has been performed with a high fertility results (i.e., high PR, FR and litter sizes). Recently, it has been shown that a satisfactory fertility outcome was accomplished by performing IUI (doses raning from 1.5-3 billion) together with fixed-time insemination (using a correlation of WOI-Oestrus duration-Ovulation time). This strategy may improve fertility of frozen boar semen when one performs insemination in a commercial pig farms (a field trial).

13. Conclusion

Based on above review, the conclusion can be drawn as follows: (I) some antioxidants, such as, Oryzanol, L-cysteine and its combination with DHA from fish oil, Vitamin E, Vitamin C, can be used in order to improve the quality of frozen boar semen" (II) the artificial insemination (i.e. timing, dose and AI techniques) by using frozen boar semen on pig farm need further investigations.

14. References

Agarwal, A., Saleh, R.A. and Bedaiwy, M.A. 2003. Role of reactive oxygen species in the pathophysiology of human reproduction. Fertil. Steril. 79:829-843.

Almlid, T. and Hofmo, P.O. 1996. A brief review of frozen semen application under Norwegian AI service conditions. Reprod. Dom. Anim. 31:169-173.

Almlid, T. and Johnson, L.A. 1988. Effects of glycerol concentration, equilibration time and temperature of glycerol addition on post-thaw viability of boar spermatozoa frozen in straws. J. Ani. Sci. 66:2899-2905.

Alvarez, J.G. and Storey, B.T. 1992. Evidence for increased lipid peroxidative damage and loss of superoxide dismutase activity as a mode of sublethal cryodamage to human sperm during cryopreservation. J. Androl. 13:232-241.

Axnér, E., Hermansson, U. and Linde-Forsberg, C. 2004. The effect of Equex STM paste and sperm morphology on frozen-thawed survival of cat epididymal sperm. Anim. Reprod. Sci. 84: 179–91.

Bailey, J.L., Lessard, C., Jacques, J., Breque, C., Dobrinski, I., Zeng, W. and Galantino-Homer, H.L. 2008. Cryopreservation of boar semen and its future importance to the industry. Theriogenology. 70:1251-1259.

Ball, B.A. and Vo, A. 2001. Osmotic tolerance of equine spermatozoa and the effects of soluble cryoprotectants on equine sperm motility, viability, and mitochondrial membrane potential. J. Androl. 22:1061-1069.

Barrios, B., Perez-Pe, R., Gallego, M., Tato, A., Osada, J., Muino-Blanco, T. and Cebrian-Perez, J.A. 2000. Seminal plasma proteins revert the cold shock damage on ram sperm membrane Biol. Reprod. 63:1531-1537.

Baumber, J., Ball, B.A., Gravance, C.G., Medina, V. and Davies-Morel, M.C. 2000. The effect of reactive oxygen species on equine sperm motility, viability, acrosomal integrity, mitochondrial membrane potential, and membrane lipid peroxidation. J. Androl. 21:621-628.

Baumber, J., Ball, B.A., Linfor, J.J. and Meyers, S.A. 2003. Reactive oxygen species and cryopreservation promote DNA fragmentation in equine spermatozoa. J. Androl. 24:895-902.

Beardon, H.J. and Fuquay, J.W. 1997. Semen evaluation. Appl Anim Rep 4th ed. 158-170.

Beconi, M.T., Francia, C.R., Mora, N.G. and Affranchino, M.A. 1993. Effect of natural antioxidants on frozen bovine semen preservation. Theriogenology. 40:841–851.

Berger, B. and Fischerleitner, F. 1992. On deep freezing of boar semen: investigations on the effects of different straw volumes, methods of freezing and thawing extenders. Reprod. Dom. Anim. 27:266-270.

Bilodeau, J. F., Blanchette, S., Gagnon, C. and Sirard, M.A. 2001. Thiols prevent H_2O_2-mediated loss of sperm motility in cryopreserved bull semen. Theriogenology. 56:275-286.

Blesbois, E., Douard, V., Germain, M., Boniface, P. and Pellet, F. 2004. Effects of n-3 polyunsaturated dietary supplementation on the reproductive capacity of male turkeys. Theriogenology. 61:537-549.

Breininger, E., Beorlegui, N.B., O'Flaherty, C.M. and Beconi, M.T. 2005. Alpha-tocopherol improves biochemical and dynamic parameters in cryopreserved boar semen. Theriogenology. 63:2126-2135.

Brzezińska-Ślebodzińska, E., Slebodziński, A.B., Pietras, B. and Wieczorek, G. 1995. Antioxidant effect of vitamin E and glutathione on lipid peroxidation in boar semen plasma. Biol. Trace. Elem. Res. 47:69-74.

Buranaamnuay, K., Tummaruk, P. and Techakumphu, M. 2010. Intra-uterine insemination using frozen-thawed boar semen in spontaneous and induced ovulating sows under field conditions. The 36th Annual Conference of the IETS/the 23rd Annual Meeting SBTE. 9-12 January 2010. Cordoba, Argentina

Buranaamnuay, K., Tummaruk, P., Singlor, J., Rodriguez-Martinez, H. and Techakumphu, M. 2009. Effects of straw volume and Equex-STM® on boar sperm quality after cryopreservation. Reprod. Dom. Anim. 44:69-73.

Bwanga, C.O., Einarsson, S. and Rodriguez-Martinez, H. 1991. Deep freezing of boar semen packaged in plastic bags and straws. Reprod. Dom. Anim. 26:117-125.

Cerolini, S., Maldjian, A., Surai, P. and Noble, R. 2000. Viability, susceptibility to peroxidation and fatty acid composition of boar semen during liquid storage. Anim. Reprod Sci. 58:99-111.

Chanapiwat, P., Kaeoket, K. and Tummaruk, P. 2009. Effects of DHA-enriched hen egg yolk and L-cysteine supplementation on quality of cryopreserved boar semen. Asian J Androl. 11: 600-608.

Chanapiwat, P., Kaeoket, K. and Tummaruk, P., 2010. The sperm DNA damage aftercryopreservation of boar semen in relation to post-thawed semen qualities, antioxidant supplementation and boars effects. Thai. J. Vet. Med. 40, 187-193.

Chanapiwat, P., Kaeoket, K. and Tummaruk, P., 2011. Cryopreservation of boar semen by egg-yolk based extenders containing lactose or fructose is better than sorbitol. J. Vet. Med. Sci. In press.

Cheng, F.P, Fazeli, A., Voorhout, W.F., Marks, A., Bevers, M.M. and Colenbrander, B. 1996. Use of peanut agglutinin to assess the acrosomal status and the zona pellucida-induced acrosome reaction in stallion spermatozoa. J Androl 1996; 17: 674–82.

Crabo, B. and Einarsson, S. 1971. Fertility of deep frozen boar spermatozoa. Acta. Vet. Scand. 12:125-127.

Curry, M.R. and Watson, P.F. 1994. Osmotic effects on ram and human sperm membranes in relation to thawing injury. Cryobiology. 31: 39-46.

Donoghue, A.M. and Donoghue, D.J. 1997. Effects of water and lipid-soluble antioxidants on turkey sperm viability, membrane integrity, and motility during liquid storage. Poult. Sci. 76:1440–1455.

Dott, H.M. and Foster, G.C.1972. A technique for studying the morphology of mammalian spermatozoa which are eosinophilic in a differential live/dead stain. J. Reprod. Fert. 29:443-445.

Eriksson, B.M. and Rodriguez-Martinez, H. 2000. Effect of freezing and thawing rate on the post-thaw viability of boar spermatozoa frozen in FlatPack and Maxi-straws. Anim. Reprod. Sci. 63: 205-220.

Eriksson, B.M., Petersson, H. and Rodriguez-Martinez, H. 2002. Field fertility with exported boar semen frozen in the new flatpack container. Theriogenology. 58: 1065-1079.

Eriksson, B.M., Vazquez, J.M. and Martinez, E.A. 2001. Effects of holding time during cooling and of type of package on plasma membrane integrity, motility and in vitro oocyte penetration ability of frozen-thawed boar spermatozoa. Theriogenology. 55:1593-1605.

Fiser, P., Fairfull, R.W., Hansen, C., Panich, P.L., Shrestha, J.N.B. and Underhill, L. 1993. The effect of warming velocity on motility and acrosomal integrity of boar sperm as influenced by the rate of freezing and glycerol level. Mol. Reprod. Dev. 34:190-195.

Fiser, P.S. and Fairfull, R.W. 1990. Combined effect of glycerol concentration and cooling velocity on motility and acrosomal integrity of boar spermatozoa frozen in 0.5 ml straws. Mol. Reprod. Dev. 25: 123-129.

Funahashi, H. and Sano, T. 2005. Select antioxidants improve the function of extended boar semen stored at 10 degrees C. Theriogenology. 63:1605-1616.

Gadea, J., Selles, E., Marco, M.A., Coy, P., Matas, C., Romar, R. and Ruiz, S. 2004. Decrease in gluthathione content in boar sperm after cryopreservation, effect of the addition of reduced glutathione to the freezing and thawing extender. Theriogenology. 62:690-701.

Garcia, J.C., Dominguez, J.C., Pena, F.J., Alegre, B., Gonzalez., Castro, M.J., Habing, G.G. and Kirkwood, R.N. 2009. Thawing boar semen in the presence of seminal plasm: effects on sperm quality and fertility. Anim. Reprod. Sci. (In press)

Garner, D.L. and Johnson, L.A. 1995. Viability assessment of mammalian sperm using SYBR-14 and propidium iodide. Biol. Reprod. 53:276-284.

Gerrits, R.J., Lunney, J.K., Johnson, L.A., Pursel, V.G., Kraeling, R.R., Rohrer, G.A. and Dobrinsky, J.R. 2005. Perspectives for artificial insemination and genomics to improve global swine populations. Theriogenology. 63:283-299.

Gilmore, J.A., Junying, D., Jun, T., Peter, A.T. and Crister, J.K. 1996. Osmotic properties of boar spermatozoa and their relevance to cryopreservation. J. Reprod. Fertil. 107:87-95.

Green, C.E. and Watson, P.F. 2001. Comparison of the capacitation-like state of cooled boar spermatozoa with true capacitation. Reproduction. 122:889-898.

Hernández, M., Roca, J., Calvete, J.J., Sanz, L., Muiño-Blanco, T., Cebrián-Pérez, J.A., Vázquez, J.M. and Martínez, E.A. 2007. Cryosurvival and in vitro fertilizing capacity post thaw is improved when boar spermatozoa are frozen in the presence of seminal plasma from good freezer boars. J. Androl. 28:689-697.

Holt, W.V. 2000a. Fundamental aspects of sperm cryobiology: the importance of species and individual differences. Theriogenology. 53:47-58.

Holt, W.V. 2000b. Basic aspects of frozen storage of semen. Anim. Reprod. Sci. 62:3–22.

Hu, J.H., Li, Q.W., Jiang, Z.L., Yang, H., Zhang, S.S. and Zhao, H.W. 2009. The cryoprotective effect of trehalose supplementation on boar spermatozoa quality. Anim. Reprod. Sci. (In press).

Iguer-Ouada, M. and Verstegen, J.P. 2001. Evaluation of the Hamilton thorne computer-based automated system for dog semen analysis. Theriogenology. 55: 733-749.

Johnson, L. A. 1985. Fertility results using frozen boar spermatozoa: 1970 to 1985. In L. A. Johnson, K. Larsson (Ed.), Deep freezing of boar semen. Uppsala, Swedish Univ Agric Sci: 199-222.

Johnson, L.A. 1998. Current developments in swine semen: preservation, artificial insemination and sperm sexing. Proceedings 15th International Pig Vet Society, Birmingham, UK, 1:225-229.

Johnson, L.A., Weitze, K.F., Fiser, P. and Maxwell, W.M.C. 2000. Storage of boar semen. Anim. Reprod. Sci. 62:142-172.

Kaeoket, K. and Tummaruk, P., 2002. Seminal plasma: its function and importance for pig artificial insemination. Thai J. Vet. Med. 32, 15-25.

Kaeoket, K., Chanapai, P., Chanchaiyaphoom, P. and Chanapiwat, P. 2011. Effect of using long term and short term extenders during cooling process on the quality of frozen boar semen. Thai. J. Vet. Med. 41, 283-288.

Kaeoket, K., Chanapiwat, P., Tummaruk, P. and Techakumphu, M. 2010. Supplementaleffect of varying L-cysteine concentrations on quality of cryopreserved boar semen. Asian. J. Androl. 12, 260-265.

Kaeoket, K., Chanapiwat, P., Tummaruk, P., Kunavongkrit, A. and Techakumphu, M., 2011. A preliminary study on using autologous and heterologous boar sperm supernatant from freezing processes as post-thawing solution: its effect on sperm motility. Trop. Anim. Health. Prod. 43, 1049-1055.

Kaeoket, K., Chanapiwat, P., Wongtawan, T. and Kunavongkrit, A. 2010. Successful intrauterine insemination (IUI) with frozen boar semen: effect of dose, volume and fixed time–AI on fertility. Thai. J. Agri. Sci. 43, 31-37.

Kaeoket, K., Persson, E. and Dalin, A.M., 2002. The influence of pre- and post-ovulatory insemination on sperm distribution in the oviduct, accessory sperm to the zona pellucida, fertilisation rate and embryo development in sows. *Anim. Reprod. Sci.* 71, 239-248.

Kaeoket, K., Sang-urai, P., Thamniyom, A., Chanapiwat, P. and Techakumphu, M. 2010. Effect of Decosahexaenoic acid (DHA) on quality of cryopreserved boar semen in different breeds. Reprod. Dom. Anim. 45, 458-463.

Kaeoket, K., Tantasuparuk, W. and Kunavongkrit, A., 2005. The effect of post-ovulatory insemination on the subsequent oestrous cycle length, embryonic loss and vaginal discharge in sows. Reprod. Domest. Anim. 40, 492-494.

Kaeoket, K., Tantiparinyakul, K., Kladkaew, W., Chanapiwat, P. and Techakumphu M. 2008. Effect of different antioxidants on quality of cryo-preserved boar semen in different breeds. Thai. J. Agri. Sci. 41, 1-9.

Kaneto, M., Harayama, H., Miyake, M. and Kato, S. 2002. Capacitation-like alterations in cooled boar spermatozoa: assessment by the chlortetracycline staining assay and immunodetection of tyrosine-phosphorylated sperm proteins Anim. Reprod. Sci. 73:197-209.

Kasetrtut, C. and Kaeoket, K. 2010. Effect of using supernatant for post-thawing solution and semen extender prior to insemination on sow reproductive performance. Thai. J. Vet. Med. 40, 171-178.

Kunavongkrit, A., Sang-Gasanee, K., Phumratanaprapin, C., Tantasuparuk, W. and Einarsson S. 2003. A study on the number of recovered spermatozoa in the uterine horns and oviducts of gilts, after fractionated or non-fractionated insemination. J. Vet. Med. Sci. 65:63-67.

Larsson, K. and Einarsson, S. 1976. Fertility of deep frozen boar spermatozoa. Acta vet. Scand. 17:43-62.

Leibo, S.P. and Songsasen, N. 2002. Cryopreservation of gametes and embryos of non-domestic species. Theriogenology. 57:303-326.

Liu, Z. and Foote, R.H. 1998. Bull sperm motility and membrane integrity in media varying in osmolality. J. Dairy. Sci. 81:1868-1873.

Maldjian, A., Pizzi, F., Gliozzi, T., Cerolini, S., Penny, P. and Noble, R. 2005. Changes in sperm quality and lipid composition during cryopreservation of boar semen. Theriogenology. 63:411-421.

Martin, M.J., Edgerton, S. and Wiseman, B. 2000. Frozen semen: a breeding protocol that results in high fecundity. Swine. Health. Prod. 8:275-277.

Martinez, E.A., Vazquez, J.M., Roca, J., Lucas, X., Gil, M.A., Parrilla, I., Vazquez, J.L. and Day, B.N. 2001. Successful non-surgical deep intrauterine insemination with small number of spermatozoa in sows. Reproduction 122, 289–296.

Maxwell, W.M.C. and Johnson, L.A. 1999. Physiology of spermatozoa at high dilution rates: the influence of seminal plasma. Theriogenology. 52:1353-1362.

Maxwell, W.M.C., Welch, G.R. and Johnson, L.A. 1997.Viability and membrane integrity of spermatozoa after dilution and flow cytometric sorting in the presence or absence of seminal plasma. Reprod. Fertil. Dev. 8:1165-1178.

Mazur, P. 1970. Cryobiology: the freezing of biological systems. Science. 168:939-949

Medeiros, C.M.O., Forell, F., Oliveira, A.T.D. and Rodrigues, J.L. 2002. Current status of sperm cryopreservation: why isn't it better? Theriogenology. 57:327-44

Moce, E., Purdy, P.H. and Graham, J.K. 2009. Treating ram sperm with cholesterol-loaded cyclodextrins improves cryosurvival. Anim. Reprod.Sci. (In press)

Nissen, A. K., Soede, N. M., Hyttel, P., Schmidt, M. and D Hoore, L. 1997. The influence of time of insemination relative to time of ovulation and farrowing frequency and litter size in sows, as investigated by ultrasonography. Theriogenology. 47: 1571-1582.

Okazaki, T., Abe, S., Yoshida, S. and Shimada, M. 2009. Seminal plasma damages sperm during cryopreservation, but its presence during thawing improves semen quality and conception rates in boars with poor post thaw semen quality. Theriogenology. 71:491–498.

Parks, J.E. and Graham, J.K. 1992. Effects of cryopreservation procedures on sperm membranes. Theriogenology. 38:209-222.

Penã, F. J., Johannisson, A., Wallgren, M. and Rodriguez Martinez, H. 2003. Antioxidant supplementation in vitro improves boar sperm motility and mitochondrial membrane potential after cryopreservation of different fractions of the ejaculate. Anim. Reprod. Sci. 78:85-98.

Perez-Llano, B., Lorenzo, J.L., Yenes, P., Trejo, A. and Garcia-Casado, P. 2001. A short hypoosmotic swelling test for the prediction of boar sperm fertility. Theriogenology. 56:387-398.

Polge, C. 1956. Artificial insemination in pigs. Vet. Rec. 68: 62-76.

Polge, C., Salamon, S. and Wilmut, I.1970 Fertilizing capacity of frozen boar semen following surgical insemination. Vet. Rec. 87:424-429.

Pukazhenthi, B., Noiles, E., Pelican, K., Donoghue, A., Wildt, D.E. and Howard, J.G. 2000. Osmotic effects on feline spermatozoa from normospermic versus teratospermic donors. Cryobiology. 40:139-150.

Purdy, P.H. and Graham, J.K. 2004 Effect of adding cholesterol to bull sperm membranes on sperm capacitation, the acrosome reaction and fertility. Biol. Reprod. 71:522-527.

Pursel, V. G., Johnson, L. A. and Schulman, L. L.1973. Effect of dilution, seminal plasma and incubation period on cold shock susceptibility of boar spermatozoa. J. Anim. Sci. 37: 528-531

Pursel, V.G., Johnson, L.A. and Schulman, L.L. 1972. Interaction of extender composition and incubation period on cold shock susceptibility of boar spermatozoa. J. Anim. Sci. 35:580-584.

Pursel, V.G., Schulman, L.L. and Johnson, L.A. 1978a. Effect of glycerol concentration on frozen boar sperm Theriogenology. 9:305-312

Pursel, V.G., Schulman, L.L. and Johnson, L.A. 1978b. Effect of Orvus ES Paste on acrosome morphology, motility and fertilizing capacity of frozen-thawed boar sperm J. Anim. Sci. 47:198-202.

Roca, J., Gil, M.A., Hernandez, M., Parrilla, I., Vazquez, J.M. and Martinez, E.A. 2004. Survival and fertility of boar spermatozoa after freeze-thawing in extender supplemented with butylated hydroxytoluene. J. Androl. 25:397-405.

Roca, J., Rodriguez, J.M., Gil, M.A., Carvajal, G., Garcia, E.M., Cuello, C., Vazquez, J.M. and Martinez, E.A. 2005. Survival and in vitro fertility of boar spermatozoa frozen in the presence of superoxide dismutase and/or catalase. J. Androl. 26:15-24.

Rodriguez-Martinez, H. 2009. Freezing of boar semen:state of the art. Proceedings 4th Asian Pig Vet Society, Tsukuba, Japan. 55-58.

Rooke, J.A., Shao, C.C. and Speake, B.K. 2001. Effects of feeding tuna oil on the lipid composition of pig spermatozoa and in vitro characteristics of semen. Reproduction. 121:315-322.

Rozeboom, K.J., Troedsson, M.H.T., Hodson, H.H., Shurson, G.C. and Crabo, B. 2000. The importance of seminal plasma on the fertility of subsequent artificial inseminations in swine J. Anim. Sci. 78:443-448

Saleh, R.A. and Agarwal, A. 2002. Oxidative stress and male infertility: From research bench to clinical practice. J. Androl. 23:737-749

Satorre, M.M., Breininger, E., Beconi, M.T. and Beorlegui, N.B. 2007. Alpha-tocopherol modifies tyrosine phosphorylation and capacitation-like state of cryopreserved porcine sperm. Theriogenolgy. 68:958-965.

Selles, E., Gadea, J., Romar, R., Matas, C. and Ruiz, S. 2003. Analysis of in vitro fertilizing capacity to evaluate the freezing procedures of boar semen and to predict the subsequent fertility. Reprod. Dom. Anim. 38:66-72.

Sharma R.K. and Agarwal, A. 1996. Role of reactive oxygen species in male infertility (Review). Urology. 48:835-850.

Sikka, S.C., Rajasekaran, M. and Hellstrom, W.J. 1995. Role of oxidative stress and antioxidants in male infertility. J. Androl. 16:464-468.

Silva, P.F.N. 2006. Physiology of peroxidation process in mammalian sperm. PhD Thesis. Utrecht University Ridderprint, Ridderkerk, 4-45.

Soede, N. M, and Kemp, B. 1997. Expression of oestrus and timing of ovulation in pigs. J Reprod Fertil (Suppl.) 52: 91-103.

Soede, N. M., Wetzels, C. C., Zondag, W., de Koning, M. A. and Kemp, B. 1995. Effects of time of insemination relative to ovulation, as determined by ultrasonography, on fertilization rate and accessory sperm count in sows. J. Reprod. Fertil. 104: 99-106.

Songsasen, N., Yu, I., Murton, S., Paccamonti, D.L., Eilts, B.E., Godke, R.A. and Leibo, S.P. 2002. Osmotic sensitivity of canine spermatozoa. Cryobiology. 44:79-90.

Strzezek, J., Lapkiewicz, S. and Lecewicz, M. 1999. A note on antioxidant capacity of boar seminal plasma. Anim. Sci. P. 17:181-188

Suzuki, K., Asano, A., Eriksson, B.M., Niwa, K., Nagai, T. and Rodriguez-Martinez, H. 2002. Capacitation status and in vitro fertility of boar spermatozoa: effects of seminal plasma, cumulus-oocyte-complexes-conditioned medium and hyaluronan Int. J. Androl. 25:84-93

Szczesniak-Fabianczyk, B., Bochenek, M., Smorag, Z. and Silvestre, M.A. 2003. Effects of antioxidants added to boar semen extender on the semen survival and sperm chromatin structure. Reprod. Biol. 3:81-87.

Tardif, S., Dube, C., Chevalier, S. and Bailey, J.L. 2001. Capacitation is associated with tyrosine phosphorylation and tyrosine kinase-like activity of pig sperm proteins. Biol. Reprod. 65: 784-792.

Thurston, L.M., Siggins, K., Mileham, A.J., Watson, P.F. and Holt, W. 2002 Identification of amplified restriction fragment length polymorphism markers linked to genes controlling boar sperm viability following cryopreservation. Biol Reprod. 2002 66:545-54.

Thuwanut, P., Chatdarong, K., Techakumphu, M. and Axnér, E. 2008. The effect of antioxidants on motility, viability, acrosome integrity and DNA integrity of frozen-thawed epididymal cat spermatozoa. Theriogenology. 70:233–240.

Tummaruk, P., Sumransap, P., Techakumphu, M. and Kunavongkrit, A. 2007. Distribution of spermatozoa and embryos in the female reproductive tract after unilateral deep intra uterine insemination in the pig. Reprod. Dom. Anim. 42:603-609.

Uysal, O. and Bucak, M.N. 2007. Effects of Oxidized Glutathione, Bovine Serum Albumin, Cysteine and Lycopene on the Quality of Frozen-Thawed Ram Semen. Acta. Vet. Brno. 76:383-390.

Vadnais, M.L. and Roberts, K.P. 2007. Effects of seminal plasma on cooling induced capacitative changes in boar sperm. J. Androl. 28. 416–422.

Vadnais, M.L., Kirkwood, R.N., Sprecher, D.J. and Chou, K. 2005b. Effect of extender, incubation temperature, and added seminal plasma on capacitation of cryopreserved, thawed boar sperm as determined by chlortetracycline staining. Anim. Reprod. Sci. 90. 347–354.

Vadnais, M.L., Kirkwood, R.N., Tempelman, R.J., Sprecher, D.J. and Chou, K. 2005a. Effect of cooling and seminal plasma on the capacitation status of fresh boar sperm as determined using chlortetracycline assay. Anim. Reprod. Sci. 87. 121-132.

Waberski, D., Sudhoff, H., Hahn, T., Jungblut, P.W., Kallweit, E., Calvete, J.J., Ensslin, M., Hoppen, H.-O., Wintergalen, N., Weitze, K.F. and Topfer-Petersen, E. 1995. Advanced ovulation in gilts by the intrauterine application of a low molecular mass pronase-sensitive fraction of boar seminal plasma. J. Reprod. Fertil. 105:247-225.

Waberski, D., Weitze, K. F., Gleumes, T., Schwarz, M., Willmen, T. and Petzoldt, R. 1994. Effect of time of insemination relative to ovulation on fertility with liquid and frozen boar semen. Theriogenology. 42: 831-840.

Wagner, H.G. and Thibier, M. 2000. World statistics for artificial insemination in small ruminants and swine. Proc 14th ICAR, Stockholm, Sweden. vol. 2, 15:3

Watson, P. F. 2000. The causes of reduced fertility with cryopreserved semen. Anim Reprod. Sci. 60-61:481-492.

Watson, P. F. and Behan, J. R. 2002. Intra-uterine insemination of sows with reduced sperm numbers: results of a commercially based field trial. Theriogenology 57: 1683-1693.

Weitze, K. F., Wagner-Rietschel, H., Waberski, D., Richter, L. and Krieter, J. 1994. The onset of heat after weaning, heat duration, and ovulation as major factors in AI timing in sows. Reprod. Dom. Anim. 29: 433-443.

Westendorf, P., Richter, L. and Treu, H. 1975. Zur Tiefgefrierung von Ebersperma. Labor- und Besamungsergebnisse mit dem Hulsenberger Pailletten-verfahren. Dtsch. Tierarztl. Wschr, 82:261-267.

Yanagimachi R. Mammalian fertilisation. In: The Physiology of Reproduction Knobil E, 2nd ed. New York: Raven Press. 189–317.

8

Cryopreservation of Genetic Diversity in Rabbit Species (*Oryctolagus cuniculus*)

Thierry Joly, Vanessa Neto and Pascal Salvetti
Université de Lyon, France, VetAgro Sup – Isaralyon,
UPSP ICE 'Interactions between Cells and their Environment', Team Cryobio
France

1. Introduction

After the ratification of the international convention on the biodiversity in Rio, a National Cryobank was created in France in 1999 to preserve the genetic resources of domestic animals (www.cryobanque.org). Particular attention was carried on Oryctolagus cuniculus species with the extension of the national cryobanking to the rabbit (Joly et al., 1998). Nowadays, this tool is very useful for the management of animal diversity in France.

Cryopreservation corresponds of all the steps of collection and long term storage of animal populations, preserved as live cells and able to generate live animals.

Cryopreservation is not a museology action to freeze the products of the past. Contrary to that, it corresponds to the practical implementation of new biotechnologies of reproduction. It includes also all the technical means to maintain the evolutionary potential of population. So, the French National Cryobank was created to secure the biological material stored at -196°C in liquid nitrogen. It constitutes a real tool to serve all stakeholders to manage the animal diversity as part of the National Charter supported by the Genetic Resources Office (www.fondationbiodiversite.fr).

2. Genetic diversity in rabbit species

The archaeological origins of rabbit are located in Spain (Bolet et al., 2000). Actually, rabbit is widely spread all around the world and can be considered according situations as wild animal, domestic animal, pets and laboratory models. The rabbit populations can be classified into 3 categories according to their genetic originality, their specific uses and the motivations of breeders involved in the *in situ* management of populations (figure 1).

2.1 "Type I" material

Type I regroups the breeds identified according to an official standard, as well as for large breeds (Butterfly, Champagne Argente, Fauve de Bourgogne...) and endangered breeds with less than 100 females (Brun Marron de Lorraine ...). These breeds are reared by fancy breeders and animals are presented regularly in local or regional meetings supported by FFC the federation of French fancy rabbit breeders (www.ffc.asso.fr). These breeds must be preserved for their patrimonial values and socio cultural interests.

2.2 "Type II" material

Type II concerns animals with one or more specific character:

- Animals carrying an identified gene (The recessive gene Sam in "jumper Alfort rabbit" evidenced by a modified walking on anterior legs; new transgenic lines recently created)
- Animals showing a specific combination of genes (histocompatible lines; allotypic strains from Basel having used as model for histocompatibility major complex studies)
- Animals with a particular genotype from a high selected population but not from the actual breeding schemes (black Orylag rabbit and Rex albinos presenting colours differing from the selected population, the divergent strain Inra 1029 made of two foetal mortality rate-diverging lines).

Most of Type II populations mainly have a scientific interest with high experimental values. Some populations could be promoted in biomedical research and by the pharmaceutical industry.

2.3 "Type III" material

Type III regroups all the populations of rabbits selected for meat production (female lines and male lines) or fur production with high economic value.

Some strains are today completely extinct and subsist only under embryo-frozen form waiting for a possible perspective of reuse or selection scheme reorientation (Dutch hymalayan, Orylag, INRA 1077 female strains...).

Other populations are still selected and the frozen biologic material represents a selection control to measure genetic progress precisely. Moreover, for security, several populations commercially spread are subject to regular cryoconservation in order to save the selection core from a sanitary risk (Inra 2066, commercial lines of Hypharm and Hycole societies...)

| Rabbit « Fauve de Bourgogne » (type I) adapted to a traditionnal breeding system | Rabbit « Brun Marron de Lorraine » (type I), endangered breed | Rabbit « Orylag® castor (type III) bred for its "Rex" fur | Rabbit "Sauteurs d'Alfort" (type II). They move on forlegs when they are stressed |

Fig. 1. Examples of different categories of rabbits

3. Principal ways of cryoconservation of rabbit genetic resources

To keep rabbit genetic resources, we can distinguish three principal ways according to the nature of frozen biological material (Figure 2).

3.1 Germinal cells

Based on the conservation of germinal cells, this way helps to conserve the gene pool of particular male or female individuals.

Mature spermatozoids (sperm collected in artificial vaginas) or immature (epididymal spermatozoids, spermatogonies in gonadic tissue) helps to save the male line. The freezing of an individual's semen (n generation) helps to obtain progeny (n+1 generation) after thawing and artificial insemination (AI) of females.

Semen samples of bad quality can be promoted by new technologies of *in vitro* fertilization (FIV) or intracytoplasmic sperm injection (ICSI), which then permits *in vitro* embryo production and progeny delivering after embryo transfer in synchronized recipient female (Daniel *et al*, 2007). Nowadays, semen freezing is not a reliable or repeatable method yet, and results after AI with rabbit frozen semen are still too inconstant to plan a routine utilization of this technique (Vicente *et al*, 1996; Moce *et al*, 2003). Only half of the sampled males can produce semen with freezable quality and approximately 50% of females give birth after insemination of thawed semen with a large variability [15%~80%]. However, this is the only available method to preserve precious males semen, mainly for type I and II.

Mature oocytes (picked up in oviduct 15 hours after ovulation) or immature oocytes (present in follicles of ovarian tissue) permits to save genetic resources by female way and to preserve cytoplasmic heredity. Mature oocytes freezing of a female individu (n generation) would permit to obtain progenies (n+1 generation) after FIV or ICSI, but no young rabbit has been obtained from thawed oocytes yet (Salvetti *et al*, 2010). But, recently, young rabbits obtained from frozen ovarian tissue are born from females transplanted by orthotopic autograft (Almodin *et al*., 2004; Neto *et al*., 2007). In emergency situations (sanitary problems, injured animals), ovarian cortex freezing, even if this method is not yet completely under control (Neto *et al*., 2008), can be proposed to save the heredity pool of an important female of type II.

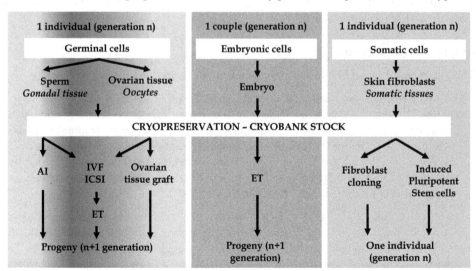

Fig. 2. Biotechnologies of reproduction applied to the cryopreservation of the rabbit genetic resources

3.2 Embryonic cells

Embryo freezing, allowing the preservation of both male and female ways, is the main way to recreate quickly a population from thawed embryo transfer. Most of the time, embryos are frozen at compacted morula stage (65h~72h *post coïtum*) even if freezing at earlier stage (4 cells stage) is possible. They can be produced *in vivo* from superovulated females or not, or *in vitro* after In Vitro Fertilization and *in vitro* culture. Embryos freezing from a planned mating (at n generation) permits to obtain progenies (at n+1 generation) after thawing and transfer in recipient females.

3.3 Somatic cells

This way has the main advantage of being easy to do, which permits to sample tissues quickly and simply on a large number of individuals. But the main difficulty is the recreation of animals after somatic cloning, still not well controlled. The cells can be isolated from different tissues: skin, cartilage, bones, blood… These cells, present in a large number in each organism, would allow the reconstitution of an individual of the same generation than the tissues giver after somatic cloning (Chesne *et al.*, 2002).

More recently, a new opportunity came with induced pluripotent stem cells, which are able to differentiate in many tissues. These cells are potential vectors for genetic traits transmission and could be used after thawing as an important core source to regenerate many individuals after nuclear transfer (Honda *et al*, 2010).

4. Frozen embryo, the main way for rabbit cryoconservation

The embryo is the favored biological material to conserve most of rabbit populations, by both male and female ways, and then preserve cytoplasmic heredity. This method has been widely proved for 15 years with more than 32 000 produced embryos but can only be applied to living and fertile animals.

4.1 Embryos production

As rabbit is naturally prolific species, two methods can be used to produce embryos:

- Without ovarian stimulation: one single injection of a GnRH analog (gonadotropic hormon) induces females ovulation at the same time as artificial insemination. Sometimes, an injection of 20-30 UI of eCG (equine chorionic gonadotrophin) three days before mating increases the females receptivity, especially the sexual resting ones;
- With ovarian stimulation by superovulation: a three days treatment of FSH injection stimulates the ovarian activity and increases the number of produced embryos per donor female (Kauffman *et al*, 1998; Salvetti *et al.*, 2007).

Overall, 73% of the treated and collected females are embryos donors and produce an average of 9.2 embryos without ovarian stimulation or with a slight eCG stimulation, and two to three times more (21.2 embryos) after a superovulation treatment (**table 1**). The embryo recovering rates (number of collected embryos/number of corpus luteum) are between 70% and 80% and those of freezable embryos (number of frozen embryos/number of collected embryos) are from 75% to 85%.

Treatment	Nb of treated females	% donor	Nb of frozen embryos	Nb of embryos per donor
No ovarian stimulation	1226	77%	8633	9.2
Superovulation (FSH)	1589	71%	23833	21.2
Total	2815	73%	32466	

Table 1. Method of embryo production (activity from 1998 to 2011)

4.2 Embryos recovering and freezing

Slaughter the females with uterine tract washing is the most effective method to collect rabbit embryos. This method is simple and easy to implement directly on the breeding place of the animals. Cryoconservation of a population, about 40 to 50 females is feasible in one single day. The method is to keep for large population of type I and III because of systematic slaughter of females.

In addition, a new method of embryos collection by endoscopy allows renewing about four times the operation on the same female, preserving the female's integrity (Garcia et al., 1991; Besenfelder et al., 1998). A lot heavier to implement, it is reserved to rare and precious animals of type II or to small effective populations of type I.

Embryos were frozen in the same cryoprotective solution containing 1.5 M DMSO and by the same slow freezing process, even more vitrification could provide good results but this method is not totally under control (Mehaisen et al., 2006).

4.3 Embryo transfer and births

After thawing of a part of these embryos, the embryo transfer results vary according to the environmental conditions defined by the recipient female genotype and breeding conditions (table 2).

Transfer conditions	Nb of recipients	% of delivering female	Nb of thawed embryos	Nb of borned pups	Embryo developmental rate
Standard controlled	277	82%	2746	1118	41%
On Field transfer	87	56%	846	176	21%
Total	364	76%	3592	1294	36%

Table 2. Pups production after transfer of thawed embryos (activity from 1998 to 2011)

The optimal environmental conditions are defined by the recipient from a mother female line placed in a control environment in an aboveground breeding (16 hours of light per day), while uncontrolled conditions are defined by a recipient placed in an uncontrolled environment (sanitary, light, temperature) in a traditional farm's breeding conditions.

Globally, 76% of the recipients give birth to young rabbits after thawed embryo transfer with an embryo development rate of 36%. It is particularly important to control the recipient's genotype and the breeding conditions in order to guarantee the population recreation after cryoconservation.

The efficiency of the cryoconservation method has been concretely proved after transfer of embryos stored for more than 15 years in liquid nitrogen. In November 2006, 69 Brun Marron de Lorraine young rabbits got born after thawing and transfer of 134 embryos in recipients. The young rabbits born from frozen embryos of 1992 were presented to public at the international agriculture show in Paris in March 2007, in collaboration with fancy breeders and the FFC (Salvetti *et al.*, 2007).

5. The national cryobank: A saving tool of rabbit genetic resources

In March 2008, the rabbit collections presented a large genetic diversity, which could be ranged according to three types of materials previously defined. Nearly sixty of rabbit populations have been cryopreserved. More than 19.000 embryos from about 1.300 rabbit doe donor have been frozen and are currently stored in liquid nitrogen at -196°C. For all these populations, three methods of embryos production have been applied according to the situation and the physiological state of rabbit does (superovulation, eCG, simple induction of ovulation). Then, all these embryos were treated according to the same freezing protocol (Joly *et al*, 1998). All the actors working for the rabbit populations management are henceforth convinced of the interest of this tool. They have actively participated to the building of the rabbit French cryobank. Public and private selectors, associations of fancy breeders and different groups of independent breeders are regular depositor.

6. A reality

The patrimonial cryobank for rabbit species is henceforth a reality. It allows to satisfy the expectations of breeders, selectors and the actors of research and biomedical industries. This last actor takes a more and more important rules in the rabbit production.

A standard method of rabbit embryos cryopreservation has been routinely applied for 15 years. Its implementation is performed on the field with a portative controlled rate freezer. This method requires a simple stimulation of the females before the collection, and an appropriate choice of the males for the breeding combinations. However, the complete conditions are not always encountered and this method is not efficient in emergency situation (for example, when an animal is rugged or during sanitary crisis).

So, another complementary methods are studied. These new way of research includes the cryopreservation of testicular tissue and epididymal spermatozoa in the male which are dead for less than two days. It includes also the cryopreservation of the rabbit ovarian tissue. Nowadays, the freezing of ear's fibroblasts is studying to produce Induced Pluripotent Stem Cells in order to use it for chimaeric animal production as tool of regeneration of initial population without genetic drift.

7. References

Almodin, C.G., Minguetti-Camara, V.C., Meister, H., Ferreira, J.O., Franco, R.L., Cavalcante, A.A., Radaelli, M.R., Bahls, A.S., Moron, A.F., Murta, C.G. (2004). Recovery of fertility after grafting of cryopreserved germinative tissue in female rabbits following radiotherapy. *Hum Reprod,* Vol.19, No.6, pp:1287-1293,

Besenfelder, U., Strouhal, C., Brem, G. (1998). A method for endoscopic embryo collection and transfer in the rabbit. *Zentralbl Veterinarmed A,* Vol.45, No.9, pp: 577-579,

Bolet, G., Monnerot, M., Arnal, C., Arnold, J., Belle, D., Bergoglio, G., Besenfelder, U., Bosze, S., Boucher, S., Brun, J.M., Chanteloup, N., Ducourouble, M.C., Durand-Tardif, M., Esteves, P.J., Ferrand, N., Hewitt, G., Joly, T., Koehl, P.F., Laube, M., Lechevestrier, S., Lopez, M., Masoero, G., Piccinin, R., Queney, G., Saleil, G., Surridge, A., Van Der Loo, W., Vanhommerig, J., Vicente, J.S., Virag, G., Zimmermann, J.M. (2000). Evaluation and conservation of european rabbit genetic resources : first results and inferences. *World rabbit Science,* Vol. 8, pp:281-315,

Chesne, P., Adenot, P.G., Viglietta, C., Baratte, M., Boulanger, L., Renard, J.P. (2002). Cloned rabbits produced by nuclear transfer from adult somatic cells. *Nat Biotechnol,* Vol.20, pp:366-369,

Daniel, N., Chesne, P., Baratte, M., Renard, J.P. (2007). Lapins produits par injection intra-cytoplasmique de spermatozoïdes (ICSI) stockés à température ambiante. *12èmes Journées de la Recherche Cunicole,* Le Mans, France, pp.33-36,

Garcia-Ximenez, F., Vicente, J.S., Santacreu, M.A. (1991). Embryo transfer in lactating rabbit does by laparoscopy. *Anim Reprod Sci,* Vol.24, pp:343-346,

Honda, A., Hirose, M., Hatori, M., Matoba, S., Miyoshi, H., Inoue, K., Ogura, A. (2010). Generation of Induced Pluripotent Stem Cells in Rabbits : Potential experimental models for human regenerative medicine. *J. Biol. Chem.* Vol.285, pp :31362-31369,

Joly, T., Rochambeau, H. de., Renard, J.P. (1998). Etablissement d'une cryobanque d'embryons pour la conservation ex situ de la diversité génétique chez le lapin : aspects pratiques. *Genet.Sel.Evol.* Vol.30, No.1, pp:259-269,

Kauffman, R.D., Schmidt, P.M., Rall, W.F., Hoeg, J.M. (1998). Superovulation of rabbits with FSH alters *in vivo* development of vitrified morulae. *Theriogenology,* Vol. 50, pp: 1081-1092,

Mehaisen, G.M.K., Viudes-de-Castro, M.P., Vicente, J.S., Lavara, R. (2006). *In vitro* and *in vivo* viability of vitrified and non-vitrified embryos derived from eCG and FSH treatment in rabbit does. *Theriogenology,* Vol.65, pp:1279-1291,

Mocé, E., Vicente, J.S., Lavara, R. (2003). Effect of freezing-thawing protocols on the performance of semen from three rabbit lines after artificial insemination. *Theriogenology,* Vol.60, pp:115-123,

Neto, V., Joly, T., Salvetti, P., Lefranc, A.C., Corrao, N., Buff, S., Guérin, P. (2007). Ovarian tissue cryopreservation in the doe rabbit: from freezing to birth. *Cryobiology,* Vol.55, pp:344,

Neto, V., Buff, S., Lornage, J., Bottollier, B., Guerin, P., Joly, T. (2008). Effects of different freezing parameters on the morphology and viability of preantral follicles after cryopreservation of doe rabbit ovarian tissue. *Fertil. Steril.* Vol.89, No.3, pp:1348-1356,

Salvetti, P., Joly, T., Boucher, S., Hurtaud, J., Renard, JP. (2007). Viability of rabbit embryos after 15 years storage in liquid nitrogen. *44th Annual Meeting of the Society for Cryobiology CRYO 2007*, Lake Louise, Canada,

Salvetti, P., Theau-Clement, M., Beckers, J.F., Hurtaud, J., Guerin, P., Neto, V., Falieres, J., Joly, T. (2007). Effect of the luteinizing hormone on embryo production in superovulated rabbit does. *Theriogenology*, Vol.67, pp:1185–1193,

Salvetti, P., Buff, S., Afanassieff, M., Daniel, N., Guérin, P., Joly, T. (2010). Structural, metabolic and developmental evaluation of ovulated rabbit oocytes before and after cryopreservation by vitrification and slow freezing. *Theriogenology*, Vol.74, pp:847 – 855,

Vicente, J.S., Viudes-de-Castro, M.P. (1996). A sucrose-DMSO extender for freezing rabbit semen. *Reprod Nutr Dev*. Vol.36, No.5, pp:485-492.

9

Cryopreservation of Rat Sperm

Hideaki Yamashiro[1] and Eimei Sato[2]
[1] Laboratory of Animal Genetics and Reproduction, Faculty of
Agriculture, Niigata University
[2] Laboratory of Animal Reproduction, Graduate School
of Agricultural Science, Tohoku University
Japan

1. Introduction

The laboratory rat, *Rattus norvegicus*, was the first mammalian species domesticated for scientific research, which work dating back to before 1850. From this auspicious beginning, the rat has become the most widely studied experimental animal model for biomedical research (Jacob, 1999). Since the development of the first inbred rat strain by King 1909, over 500 inbred rat strains have been developed for a wide range of biochemical and physiological phenotypes and different disease models (Aitman et al., 2008, Canzian. 1997). In the last decade, there has been an extraordinary increase in rat genomic resources (Gibbs et al., 2004, Pennisi. 2004), and the advent of knock-out technology allow the insertion or deletion of individual genes into the rat by advances in stem-cell technology (Geurts et al., 2009, Izsvak et al., 2010, Tong et al., 2010). Thus, a wide array of research opportunities now open up, especially in studies involving the laboratory rat (Hamra. 2010). However, protocol for sperm cryopreservation and oocytes fertilized *in vitro* by using cryopreserved sperm are still under development for preservation of most rat strains. Therefore, greater use of the cryopreservation of rat sperm may provide an essential resource to preserve and increase the number of valuable genetic strains for research and application.

In this chapter, we will introduce several of these approaches to cryopreserving the rat sperm. It will be valuable for developing new freezing extender for cryopreservation of rat sperm, and might be applied to other reproductive technologies in this species for preservation of valuable rat strains.

2. Reporting studies of rat sperm cryopreservation

The first live-born rat derived from frozen-thawed sperm were successfully reported artificial insemination (AI) by using frozen-thawed rat sperm, and also reported cryopreservation of several strains of rat sperm, including those from mutant and transgenic rats (Nakatsukasa et al., 2001, Nakatsukasa et al., 2003). More recent publication by same group confirmed that cryopreserved rat sperm can be revitalized and result in the birth of live offspring through embryo transfers after in vitro fertilization (*IVF*) (Seita et al., 2009a). Although the authors mentioned by another publication that intracytoplasmic sperm injection (ICSI) is the only way to routinely obtain offspring routinely derived from oocytes fertilized in vitro using fresh and cryopreserved sperm (Seita et al., 2009b). Hagiwara et al, (2009) also pointed out that only one group has reported successful cryopreservation of rat sperm, subsequently used for AI and yielding live offspring. This result has not yet been repeated by other investigators and labs, and further investigation of the ability to yield viable rat sperm after cryopreservation is urgently needed. Further, each of cryopreservation procedures has not been completed for several mammalian species to date, the basic science and technology required to do so is rapidly becoming available and this should be completed for a number of species, in the future (Agca and Critser, 2002). Those of the information will also allow further improvements in cryopreservation of rat sperm from various mammalian species.

3. Cryodiluent for rat sperm

The characteristics of rodent sperm *in vitro* differ from that of other mammalian sperm largely due to differences in sperm membrane lipid content or composition (Parks and Lynch, 1992). The morphology of rodent sperm shows a longer tail when compared to that of sperm from domestic animals (Cardullo and Baltz, 1991). Rat sperm is extremely sensitive to a number of environmental changes, such as centrifugation, pH, viscosity, osmotic stress (Varisli et al., 2009a: Nakatsukasa et al., 2003: Chularatnatol, 1982: Si et al., 2006). Rat sperm have therefore proven to be more difficult to cryopreserve than other mammalian sperm, including that of the mouse, and current survival rates for sperm cryopreservation are still inadequate for AI, *IVF* and safe preservation of most rat strains. Important factors affecting sperm cryopreservation are cooling, freezing, thawing, and the composition of cryoprotectant in the freezing extender. In the rat, Nakatsukasa et al. (2001) employed a freezing medium that contained lactose monohydrate, Equex STM, and egg yolk solution. Based on the experimental conditions and extender components described in the current study, we offer the following suggestions to those attempting to cryopreserve epididymal rat sperm. Moreover, Varisli et al., (2009b) investigated that effect of chilling on the motility and acrosomal integrity of rat sperm in the presence of various extenders. They found that the addition of glycerol or propylene glycol to either Tris–citrate or TEST extender or of DMSO into lactose monohydrate, Tris–citrate, or TEST extender resulted in optimal motility rates.

4. Identified optimal energy substrates and other components of rat sperm cryodiluent

4.1 Freezability of rat sperm induced by raffinose in modified Krebs–Ringer bicarbonate (mKRB) based extender solution.

We first studied to develop an ideal freezing extender and method for rat sperm cryopreservation (Yamashiro et al., 2007). Experiments were conducted to study its post-thaw characteristics when freezing with raffinose-free buffer or various concentrations of raffinose and egg yolk dissolved in distilled and deionised water, PBS, or mKRB based extender. Different concentrations of glycerol, or Equex STM dissolved in either PBS or mKRB containing egg yolk were also tested. Based on the data from these experiments, further experiments tested how different sugars such as raffinose, trehalose, lactose, fructose, and glucose dissolved in mKRB with Equex STM and egg yolk supplementation affected the post-thaw characteristics of cryopreserved sperm. Beneficial effects on the post-thaw survival of sperm were obtained when raffinose in mKRB was used with Equex STM, and egg yolk (Fig.1).

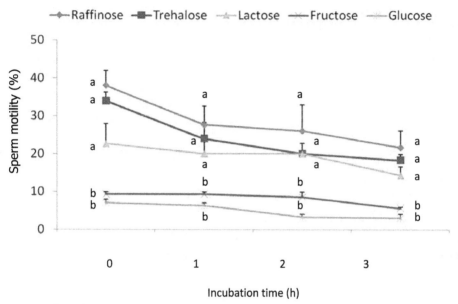

Fig. 1. Motility of frozen-thawed rat sperm which were frozen in different sugars dissolved in an mKRB egg yolk, and then subjected to a thermal resistance test at 37°C. Values are the mean ±SEM (n=3). [a-b] Different superscripts within the same column denote significant differences ($P < 0.05$).

Mammalian sperm provide energy for metabolic requirements by both mitochondrial oxidative respiration and glycolysis, and sperm motility is driven by the flagellum and is dependent on the availability of an adequate and continued supply of ATP (Cummins and Woodall, 1985). ATP is used by the dynein ATPases that function as the flagellar motors and in protein kinase A-mediated signal transduction pathways to regulate motility throughout the tail (Cao et al., 2006). In the mouse sperm, Mukai and Okuno (2004) reported that glycolysis has an important role in providing the ATP required for mouse sperm motility than mitochondrial respiration throughout the length of the flagellum. Bunch et al. (1998) also suggested that mouse sperm utilize glycolysis to generate ATP in the principal piece of the tail. While, Odet et al., (2011) demonstrated that lactate dehydrogenase (LDH) is responsible for the maintenance of energy metabolism in progressive and hyperactivated in mouse sperm. In rat sperm, Gallina et al. (1994) was studied the operation of shuttle functions for the ATP reconstituted systems in the mitochondria present in the middle place of rat, mouse and rabbit. They showed that the redox couple lactate/pyruvate and lactate dehydrogenase are active with rat and rabbit mitochondria, and it does not work with mouse. From these examining the energy metabolism of sperm, the glycolytic activity is exclusively responsible for the generation of mouse sperm metabolism, and mitochondrial metabolism seems to interacted with motility activity of rat sperm. These findings in conjunction with the present study indicate that successful cryopreservation of rat sperm in the presence of glucose, lactate and pyruvate in mKRB egg yolk extender solutions may be achieved through the ability to synthesize ATP, which could have profound effects on sperm metabolism and thereby impart a greater endurance against freeze–thawing damage.

4.2 Lactate and adenosine triphosphate in the extender enhance the cryosurvival of rat sperm

On the basis of the results of the previous experiments, we hypothesized that the metabolic state of sperm before cryopreservation would influence their survival during this stressful process (Yamashiro et al., 2010a). We evaluated the cryosurvival of rat sperm preserved in raffinose–mKRB egg yolk extender supplemented with various energy-yielding substrates (glucose, pyruvate, lactate, and ATP) and assessed the effect on sperm oxygen consumption. The incubation of sperm in lactate-free extender decreased sperm motility and oxygen consumption before and after thawing compared with those of sperm in glucose- and pyruvate-free mediums. We then focused on the effect of supplementing the extender with lactate and found that sperm frozen and thawed in extender supplemented with lactate exhibited the highest motility (Fig. 2). When we supplemented extender containing lactate with ATP, sperm frozen and thawed in the extender supplemented with ATP exhibited considerably higher motility and viability than those of sperm frozen and thawed in ATP-free extender (Fig. 3). Especially, exogenous ATP was observed that it dramatically induced the cryosurvival of rat sperm (Fig.4). Moreover, this may involve a lactate-transport system

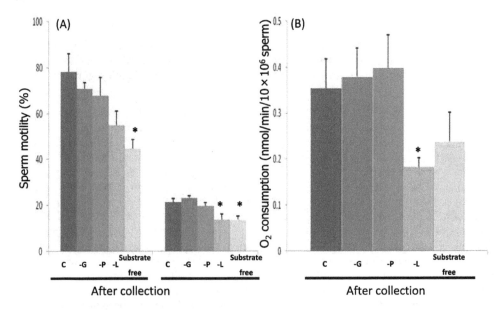

Fig. 2. Effect of the substrates in raffinose-mKRB egg yolk extender on the characteristics of after collected and frozen-thawed rat sperm (A) and oxygen consumption of after collected sperm during incubation at 37°C for 10 min (B). Indicates C; control, G; glucose, P; pyruvate, L; lactate. Values are mean ± SEM (n=5), respectively. Statistical difference (P<0.05) in comparison to the control is indicated by asterisk (*).

for regenerating cytoplasmic ATP throughout the principal piece of rat sperm in Fig.5 (Yamashiro et al., 2009). This thought is also in concert with the presence of a unique pathway that utilizes lactate and extracellular ATP in the rat sperm; this suggests new possibilities for energy production and translocation mechanisms related to motility, fertility, and freezability of rat sperm. These results provide the first evidence that supplementation of the raffinose–mKRB egg yolk extender with lactate and ATP increases of number of motile sperm before freezing and enhances the cryosurvival of rat sperm.

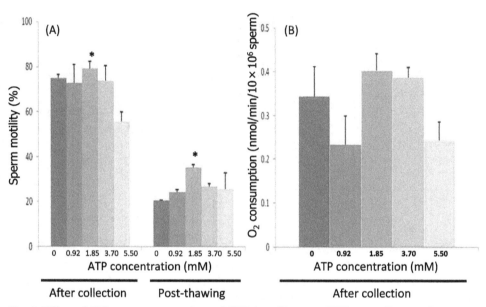

Fig. 3. Effect of different concentrations of ATP in raffinose-mKRB- gg yolk extender containing lactate on the characteristics of after collected and frozen-thawed rat sperm (A) and oxygen consumption of after collected sperm during incubation at 37°C for 10 min (B). Values are mean±SEM (n=5), respectivaly. Statistical difference ($P < 0.05$) in comparison to the control is indicated by asterisk (*).

The sperm-specific enzyme LDH isozyme C4 is located in the cytosol and the matrix of the mitochondria in the midpiece of rat sperm. Further, a study (Gallina et al., 1994) has revealed that both a shuttle involving the redox couple lactate–pyruvate and LDH C4 are active in rat sperm mitochondria. In another study (Harris et al., 2005), the lactate concentration in oviductal fluids was 10-fold higher than the glucose concentration, and the lactate concentration in the uterine fluids was 15-fold higher than the glucose concentration during the murine estrous cycle. Therefore, it is very likely that lactate is used by rat sperm as an essential substrate to maintain highly regulated ATP production and dissipation: lactate in the cytosol and mitochondrial matrix is oxidized to pyruvate by mitochondrial LDH isozyme C4, and pyruvate is oxidized through the Krebs cycle and electron transport chain (Brooks et al., 1994, Brooks. 2002, Montamat et al., 1988, Poole and Halestrap, 1993). To our knowledge, our findings are the first evidence showing that rat sperm can use exogenous lactate in the cryodiluent as an essential substrate to maintain highly regulated metabolic capacity and that this lactate acts as an energy substrate for mitochondria to the mobilization of fresh and frozen–thawed sperm.

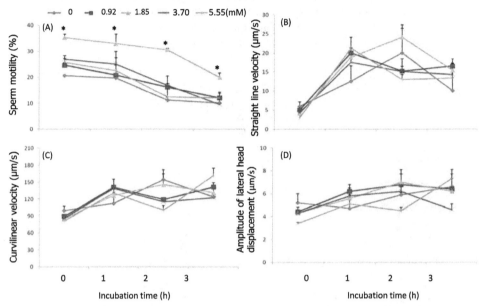

Fig. 4. Effect of different concentrations of ATP in raffinose-mKRB egg yolk medium containing lactate on the motility (A), straight line velocity (B), curvilinear velocity (C), and amplitude of lateral head displacement (D) of frozen-thawed rat sperm during incubation at 37 °C for 3 h. Values are represented as mean ± SEM (n = 3). Statistical difference (P<0.05) in comparison to the control is indicated by asterisk (*).

Mitochondria, the site of ATP generation due to oxidative phosphorylation, are localized solely in the midpiece of sperm (Millette et al., 1973). The oxidative production of ATP through the Krebs cycle is an essential function of the midpiece mitochondria for motility (Suarez et al., 2007). The mitochondrial biochemical pathways of oxidative phosphorylation are 15 times more efficient than is anaerobic glycolysis for ATP production (Cardullo and Baltz. 1991, Ruiz-Pesini et al., 2007). These findings also support our arguments that the energy production and dissipation in rat sperm are highly dependent on the mitochondria.

The present study showed that supplementation of raffinose–mKRB egg yolk extender with lactate and exogenous ATP considerably increases sperm motility before freezing, thus improving the survivability of sperm after cryopreservation. Exogenous ATP in the freezing medium may be responsible for the generation of multiple metabolic signals that appear to be related to the sperm motility through a rise in calcium levels (Gibbons.1963, Kinukawa et al., 2006, Litvin et al., 2003, Luria et al., 2002, Ren et al., 2001, Rodriguez-Miranda et al., 2007); this reaction increases de novo ATP synthesis before freezing and may contribute to the remobilization of sperm after freezing-thawing. The motility of ram sperm was restored by exogenous ATP that crossed plasma membrane when the membrane was damaged by cryopreservation (Holt et al., 1992). In light of that finding, we cannot discount that our result is caused by the facultative transport of ATP across plasma membrane because of damage during freezing, thereby allow ing substrates to directly access ATP and allowing adenosine triphosphatase to use ATP directly to generate energy for the mobilization of rat sperm.

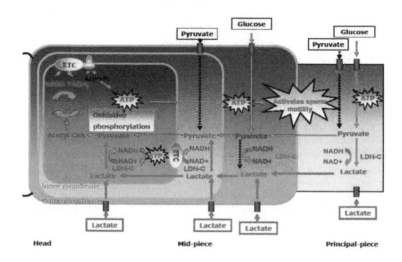

Fig. 5. Hypothesis of lactate-transport system for energy production and translocation in the rat sperm.

4.3 Extracellular ATP and dibutyryl cAMP enhance the freezability of rat sperm

The basic mechanochemical event underlying sperm motility is ATP-induced microtubule sliding (Brokaw. 1972). ATP associated with the dynein arms on outer-doublet microtubules provides the energy required for this process (Warner and Mitchell, 1980). ATP, calcium, and cAMP have received considerable attention as potential primary regulators of sperm motility in several species of animals (Aoki et al., 1996, Lindemann. 1978, Lindemann and Gibbons. 1975). Extracellular ATP acts on sperm by triggering a purinergic receptor-mediated increase in the intracellular calcium level; this increase may produce several downstream effects that enhance sperm motility (Luria et al., 2002, Rodriguez-Miranda et al., 2008). Increased calcium levels presumably activate soluble adenylyl cyclase, thereby increasing the cAMP concentration in sperm (Cook and Babcock. 1983, Garbers. 2001, Ren et al., 2001). cAMP induces protein phosphorylation by activating protein kinase A (Fujinoki et al., 2004, Kinukawa et al., 2006) and mediates calcium influx into sperm via the CatSper calcium ion channels (Cook and Babcock, 1983, Garbers. 2001, Ren et al., 2001). In addition, cAMP may elevate mitochondrial calcium levels (Degasperi et al., 2006), thereby activating the calcium-dependent dehydrogenases involved in the Krebs cycle and providing ATP required for sperm motility.

Previously, we showed that rat sperm become freezable when diluted in ATP-containing raffinose–mKRB egg yolk extender (Yamashiro et al., 2010a). This finding indicates the existence of a unique pathway that utilizes extracellular ATP in rat sperm and suggests that extracellular ATP produces several downstream effects that improve sperm motility by increasing calcium levels or by activating cAMP signal transduction pathways. Further elucidation of the role of extracellular ATP in the energy-synthetic processes and motility-regulation system of rat sperm could lead to improved motility, freezability, and fertilizing ability of the sperm. We, therefore, evaluated the freezability of rat sperm preserved in raffinose–mKRB egg yolk extender with ATP, ionomycin (a calcium ionophore), and dibutyryl cAMP (dbcAMP; a membrane-permeable cAMP analog) under various conditions. We also determined the effects of these agents on oxygen consumption by sperm. Sperm cryopreservation was considered successful if frozen–thawed sperm fertilized oocytes. To improve the effectiveness of in vitro fertilization (*IVF*), we determined whether ATP- and dbcAMP-supplemented *IVF* media improve the fertilizing ability of sperm. We also attempted artificial insemination with frozen–thawed rat sperm.

Results showed that rat sperm become freezable when diluted in ATP, and dbcAMP-containing raffinose–mKRB egg yolk extender (Yamashiro et al., 2010b). This finding indicates the existence of a unique pathway that utilizes extracellular ATP in rat sperm and suggests that extracellular ATP produces several downstream effects that improve sperm motility by increasing calcium levels or by activating cAMP signal transduction pathways. Further elucidation of the role of extracellular ATP in the energy-synthetic processes and motility-regulation system of rat sperm could lead to improved motility, freezability, and fertilizing ability of the sperm (Fig. 6). The results showed that the cryopreservation of rat

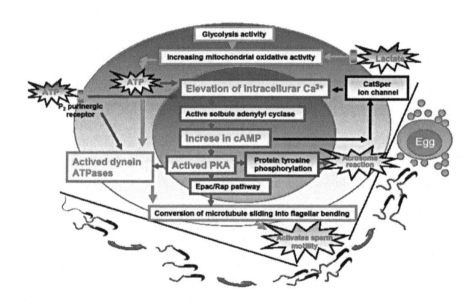

Fig. 6. Hypothesis of utilization of exogenous ATP pathway in the rat sperm.

sperm in raffinose–mKRB egg yolk extender supplemented with ATP and dbcAMP rendered sperm from rats freezable. This finding indicates that ATP- and dbcAMP-containing extenders improved the postthaw motility and fertilizing ability of cryopreserved rat sperm. Moreover, the *IVF* medium developed in the current study may be effective for the *in vitro* production of embryos from cryopreserved rat sperm. More recently, Vasudevan et al., (2011) reported that treatment of mouse sperm with extracellular ATP enhanced *IVF* rates in outbred and hybrid mice. Thus, this chapter was introduced how we identified optimal energy substrates and of rat sperm cryodiluent and *IVF* medium.

5. Concluding remarks

In this chapter described in detail the effects of the various components of cryodiluent that are used for rat sperm cryopreservation. We found that lactate, ATP and dbcAMP conferred freezability on rat sperm and enhanced oocyte fertilization by frozen-thawed sperm. In addition, if the rat sperm can be controlled flagellar movement in a way that changed by energy-yielding substrates such as lactate, it would be possible to enhance their fertilizing ability of rat sperm for *in vitro* fertilization by using a both of the fresh and frozen-thawed sperm (Fig. 7). However, there is still a lack of information for physiological trigger

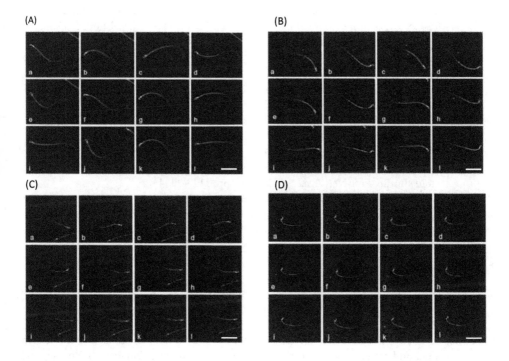

Fig. 7. Representative patterns of rat sperm movement which were extended in the presence （A) or absence （C and D) of lactate in raffinose-mKRB egg yolk solution. In B, sperm was incubated at 37°C for 3 h in the lactate containing solution. Panels a to l indicate the turn. Scale bars = 50µm

regarding how rat sperm switches on their flagellar movement at the real-time of fertilization, such as "hyper-activation" or "ultra-activation" (Yamashiro et al., 2009). Further, we believe that the *IVF* medium developed in our study is effective for the *in vitro* production of embryos from cryopreserved rat sperm. These results not only indicate that the cryopreservation of rat sperm with the present method can be applied to reproductive technologies but also indicate that exogenous lactate, rather than glucose and pyruvate, exerts a mediating effect on energy-dependent synthetic processes.

In conclusion, this is one of the very few information in the field that actually test the components of a cryodiluent in a logical manner for rat sperm cryopreservation. From this point of view, the chapter is likely to be important not only for freezing of rat sperm but also for freezing of sperm of other species in general.

6. References

Agca, Y., & Critser, J.K. (2002). Cryopreservation of spermatozoa in assisted reproduction. Semin Reprod Med. 20, pp.15–23.

Aitman, T.J., Critser, J.K., Cuppen, E., Dominiczak, A., Fernandez-Suarez, X.M., Flint, J., Gauguier, D., Geurts, A.M., Gould, M., Harris, P.C., Holmdahl, R., Hubner, N., Izsvák, Z., Jacob, H.J., Kuramoto, T., Kwitek, A.E., Marrone, A., Mashimo, T., Moreno, C., Mullins, J., Mullins, L., Olsson, T., Pravenec, M., Riley, L., Saar, K., Serikawa, T., Shull, J.D., Szpirer, C., Twigger, S.N., Voigt, B., & Worley, K. (2008). Progress and prospects in rat genetics: a community view. *Nat Genet.* 40, pp. 516–522.

Aoki, F., Sakai, S., & Kohmoto, K. (1999). Regulation of flagellar bending by cAMP and Ca^{2+} in hamster sperm. *Mol Reprod Dev.* 53, pp. 77–83.

Brokaw, C.J. (1972). Flagellar movement: a sliding filament model. *Science* 178, pp. 455–462.

Brooks, G.A., Dubouchaud, H., Brown, M., Sicurello, J.P., & Butz, C.E. (1999). Role of mitochondrial lactate dehydrogenase and lactate oxidation in the intracellular lactate shuttle. *Proc Natl Acad Sci USA.* 96, pp. 1129–1134.

Brooks. G,A. (2002). Lactate shuttles in nature. *Biochem Soc Trans.* 30, pp. 258–264.

Bunch, D.O., Welch, J.E., Magyar, P.L., Eddy, E.M., & O'Brien, D.A. (1998). Glyceraldehyde 3-phosphate dehydrogenase-S protein distribution during mouse spermatogenesis. *Biol Reprod.* 58, pp. 834–841.

Canzian, C. (1997). Phylogenetics of the laboratory rat rattus norvegicus. *Genome Res.* 7, pp. 262–267.

Cao, W., Haig-Ladewig, L., Gerton, G.L. & Moss, S.B. (2006). Adenylate kinases 1 and 2 are part of the accessory structures in the mouse sperm flagellum. *Biol Reprod.* 75, pp. 492–500.

Cardullo, R.A. & Baltz, J.M. (1991). Metabolic regulation in mammalian sperm: Mitochondrial volume determines sperm length and flagellar beat frequency. *Cell Motil Cytoskeleton.* 19, pp. 180–188.

Chularatnatol, M. (1982). Motility initiation of quiescent spermatozoa from rat caudal epididymis: effects of pH, viscosity, osmolality and inhibitors, *Int J Androl.* 5, pp. 425–436.

Cook, S.P., & Babcock, D.F. (1983). Activation of Ca²⁺ permeability by cAMP is coordinated through the pHi increase induced by speract. *J Biol Chem.* 268, pp. 22408–22413.

Cummins, J.M., & Woodall, P.F. (1986). On mammalian sperm dimensions. *J Reprod Fertil.* 75, pp. 153–175.

Degasperi, G.R., Velho, J.A., Zecchin, K.G., Souza, C.T., Velloso, L.A., Borecký, J., Castilho, R.F., & Vercesi, A.E. (2006). Role of mitochondria in the immune response to cancer: a central role for Ca²⁺. *J Bioenerg Biomembr.* 38, pp. 1–10.

Fujinoki, M., Kawamura, T., Toda, T., Ohtake, H., Ishimoda-Takag,i T., Shimizu, N., Yamaoka, S., & Okuno, M. (2004). Identification of 36 kDa phosphoprotein in fibrous sheath of hamster spermatozoa. *Comp Biochem Physiol B Biochem Mol Biol.* 137, pp. 509–520.

Gallina, F.G., Deburgos, N.M.G., Burgos, C., Coronel, C.E., & Blanco, A. (1994). The lactate-pyruvate shuttle in spermatozoa: operation in vitro. *Arch Biochem Biophys.* 308, pp. 515–519.

Garbers, D.L. (2001). Ion channels. Swimming with sperm. *Nature* 413, pp. 581–582.

Geurts, A.M., Cost, G.J., Freyvert, Y., Zeitler, B., Miller, J.C., Choi, V.M., Jenkins, S.S., Wood, A., Cui, X., Meng, X., Vincent, A., Lam, S., Michalkiewicz, M., Schilling, R., Foeckler, J., Kalloway, S., Weiler, H., Ménoret, S., Anegon, I., Davis, G.D., Zhang, L., Rebar, E.J., Gregory, P.D., Urnov, F.D., Jacob, H.J., & Buelow, R. (2009). Knockout rats via embryo microinjection of zinc-finger nucleases. *Science* 325, pp. 433.

Gibbons, I.R. (1963). Studies on the protein components of cilia from Tetrahymena pyriformis. *Proc Natl Acad Sci USA.* 50, pp. 1002–1010.

Gibbs, R.A., Weinstock, G.M., Metzker, M.L. et al., (2004). Genome sequence of the Brown Norway rat yields insights into mammalian evolution. *Nature* 428, pp. 493–521.

Hagiwara, M., Choi, J.H., Devireddy, R.V., Roberts, K.P., Wolkers, W.F., Makhlouf, A., & Bischof, J.C. (2009). Cellular biophysics during freezing of rat and mouse sperm predicts post-thaw motility. *Biol Reprod.*1, pp. 700–706

Hamra, F.K. (2010). Gene targeting: Enter the rat. *Nature* 467, pp. 211–213.

Harris, S.E., Gopichandran, N., Picton, H.M., Leese, H.J., & Orsi, N.M. (2005). Nutrient concentrations in murine follicular fluid and the female reproductive tract. *Theriogenology* 64, pp. 992–1006.

Holt, W.V., Head, M.F., & North, R.D. (1992). Freeze-induced membrane damage in ram spermatozoa is manifested after thawing: observations with experimental cryomicroscopy. *Biol Reprod.* 46, pp. 1086–1094.

Izsvak, Z., Fröhlich, J., Grabundzija, I., Shirley, J.R., Powell, H.M., Chapman, K.M., Ivics, Z., & Hamra, F.K. (2010). Generating knockout rats by transposon mutagenesis in spermatogonial stem cells. *Nat Methods.* 7, pp. 443–445.

Jacob, H.J. (1999). Functional genomic and rat model. *Genome Res.* 9, pp. 1013–1016.

Kinukawa, M., Oda, S., Shirakura, Y., Okabe, M., Ohmuro, J., Baba, S.A., Nagata, M., & Aoki, F. (2006). Roles of cAMP in regulating microtubule sliding and flagellar bending in demembranated hamster spermatozoa. *FEBS Lett.* 580, pp. 1515–1520.

Lindemann, C.B. (1978). A cAMP-induced increase in the motility of demembranated bull sperm models. *Cell* 13, pp. 9–18.

Lindemann, C.B., & Gibbons, I.R. (1975). Adenosine triphosphate-induced motility and sliding of filaments in mammalian sperm extracted with Triton X-100. *J Cell Biol.* 65, pp. 147–162.

Litvin, T.N., Kamenetsky, M., Zarifyan, A., Buck, J., & Levin, L.R. (2003). Kinetic properties of 'soluble' adenylyl cyclase: synergism between calcium and bicarbonate. *J Biol Chem.* 278, pp. 15922–15926.

Luria, A., Rubinstein, S., Lax, Y., & Breitbart, H. (2002). Extracellular adenosine triphosphate stimulates acrosomal exocytosis in bovine spermatozoa via P2 purinoceptor. *Biol Reprod.* 66, pp. 429–437.

Millette, C.F., Spear, P.G., Gall, W.E., & Edelman, G.M. (1973). Chemical dissection of mammalian spermatozoa. *J Cell Biol.* 58, pp. 662–675.

Montamat, E.E., Vermouth, N.T., & Blanco, A. (1988). Subcellular localization of branched-chain amino acid aminotransferase and lactate dehydrogenase C4 in rat and mouse spermatozoa. *Biochem J.* 255, pp. 1053–1056.

Mukai, C. & Okuno, M. (2004). Glycolysis plays a major role for adenosine triphosphate supplementation in mouse sperm flagellar movement. *Biol Reprod.* 71, pp. 540–547.

Nakatsukasa, E., Inomata, T., Ikeda, T., Shino, M., & Kashiwazaki, N. (2001). Generation of live rat offspring by intrauterine insemination with epididymal spermatozoa cryopreserved at -196 degrees C, *Reproduction* 122, pp. 463–467.

Nakatsukasa, E., Kashiwazaki, N., Takizawa, A., Shino, M., Kitada, K., Serikawa, T., Halamata, Y., Kobayashi, E., Takahashi, R., Ueda, M., Nalashima, K., & Nakagata, N. (2003). Cryopreservation of spermatozoa from closed colonies, and inbred, spontaneous mutant, and transgenic strains of rats, *Comp. Med.* 53, pp. 639–641.

Odet, F., Gabel, S.A., Williams, J., London, .RE., Goldberg, E., & Eddy, E.M. (2011). Lactate Dehydrogenase C (LDHC) and Energy Metabolism in Mouse Sperm. *Biol Reprod.* 85, pp. 556-564

Parks, J.E., & Lynch, D.V. (1992). Lipid composition and thermotropic phase behavior of boar, bull, stallion, and rooster sperm membranes, *Cryobiology* 29, pp. 255–266.

Pennisi, E. (2004). Genetics DNA reveals diatom's complexity. *Science* 306, pp. 31.

Poole, R.C., & Halestrap, A.P. (1993). Transport of lactate and other monocarboxylates across mammalian plasma membranes. *Am J Physiol.* 264(4Pt 1), pp. C761–C782.

Ren, D., Navarro, B., Perez, G., Jackson, A.C., Hsu, S.Q., Shi, Q., Tilly, J.L., & Clapham, D.E. (2001). A sperm ion channel required for sperm motility and male fertility. *Nature* 413, pp. 603–609.

Rodriguez-Miranda, E., Buffone, M.G., Edwards, S.E., Ord, T.S., Lin, K., Sammel, M.D., Gerton, G.L., Moss, S.B., & William, C.J. (2008). Extracellular adenosine 5'-triphosphate alters motility and improves the fertilizing capability of mouse sperm. *Biol Reprod.* 79, pp. 164–171.

Ruiz-Pesini, E., Díez-Sánchez, C., López-Pérez, M.J., & Enríquez, J.A. (2007). The role of the mitochondrion in sperm function: is there a place for oxidative phosphorylation or is this a purely glycolytic process? *Curr Top Dev Biol.* 77, pp. 3–19.

Seita, Y., Sugio, S., Ito, J., & Kashiwazaki, N. (2009a). Generation of live rats produced by in vitro fertilization using cryopreserved spermatozoa. *Biol Reprod.* 80, pp. 503–510.

Seita, Y., Ito, J., & Kashiwazaki, N. (2009b). Removal of acrosomal membrane from sperm head improves development of rat zygotes derived from intracytoplasmic sperm injection. *J Reprod Dev*, 2009, 55, pp. 475–479.

Si, W., Benson, J.D., Men, H., & Critser, J.K. (2006). Osmotic tolerance limits and effects of cryoprotectants on the motility, plasma membrane integrity and acrosomal integrity of rat sperm. *Cryobiology* 53, pp. 336–348.

Suarez, S.S., Marquez, B., Harris, T.P., & Schimenti, J.C. (2007). Different regulatory systems operate in the midpiece and principal piece of the mammalian sperm flagellum. *Soc Reprod Fertil Suppl.* 65, pp. 331–334.

Tong, C., Li, P., Wu, N.L., Yan, Y., & Ying, Q.L. (2010). Production of p53 gene knockout rats by homologous recombination in embryonic stem cells. *Nature* 467, pp. 211–213

Varisli, O., Uguz, C., Agca, C., & Agca, Y. (2009a). Various physical stress factors on rat sperm motility and integrity of acrosome and plasma membrane. *J Androl.* 30, pp. 75–86.

Varisli, O., Uguz, C., Agca, C., & Agca, Y. (2009b). Effect of chilling on the motility and acrosomal integrity of rat sperm in the presence of various extenders. *J Am Assoc Lab Anim Sci.* 48, pp. 499–505.

Vasudevan, K., Sztein, J.M. Treatment of sperm with extracellular adenosine 5'-triphosphate improves the in vitro fertility rate of inbred and genetically modified mice with low fertility. (2011). *Theriogenology* 76, pp. 729-736.

Warner, F.D., & Mitchell, D.R. (1980). Dynein: the mechanochemical coupling adenosine triphosphatase of microtubule-based sliding filament mechanisms. *Int Rev Cytol.* 66, pp. 1–43.

Yamashiro, H., Han, Y.J., Sugawara, A., Tomioka, I., Hoshino, Y., & Sato, E. (2007). Freezability of rat epididymal sperm induced by raffinose in modified Krebs-Ringer bicarbonate (mKRB) based extender solution. *Cryobiology* 55, pp. 285–294.

Yamashiro, H., Toyomizu, M., Kadowaki, A., Takeda, Z., Nakazato, F., Toyama, N., Kobayashi, J., & Sato, E. (2009).Oxidation of exogenous lactate by lactate dehydrogenase C in the midpiece of rat epididymal sperm is essential for motility and oxidative activity. *Am J Appl Sci.* 6, pp. 1854–1859.

Yamashiro, H., Toyomizu, M., Kikusato, M., Toyama, N., Sugimura, S., Hoshino, Y., Abe, H., Moisyadi, S., & Sato, E. (2010a). Lactate and adenosine triphosphate in the extender enhance the cryosurvival of rat epididymal sperm. *J Am Assoc Lab Anim Sci.* 49, pp. 155–161.

Yamashiro, H., Toyomizu, M., Toyama, N., Aono, N., Sakurai, M., Hiradate, Y., Yokoo, M., Moisyadi, S., & Sato, E. (2010b). Extracellular ATP and dibutyryl cAMP enhance the freezability rat epididymal sperm. *J Am Assoc Lab Anim Sci.* 49, pp. 167–172.

10

Review on Ovarian Cryopreservation in Large Animals and Non-Human Primates

Milan Milenkovic, Cesar Díaz-Garcia and Mats Brännström
Department of Obstetrics and Gynecology, Sahlgrenska Academy
University of Gothenburg
Sweden

1. Introduction

The ultimate aim of ovarian cryopreservation research is naturally to increase the effectiveness of this fertility preservation procedure in female cancer victims and much of this research on whole ovary cryopreservation, ovarian cortex cryopreservation and transplantation has been performed in animal models. However, ovarian cryopreservation could also be used in the future in programs with the purpose to rescue endangered species (Santos et al., 2010) and certain specific strains of animals (Dorsch et al., 2004). Due to the ethical barriers in the research on human tissue and the shortage of human premenopausal ovarian tissue for research purposes, there is a need to find animal models that are reasonably analogous to the human. As a general rule, animal models have to be comparable in biochemical, physiological and anatomical characteristics to the human so that the results can be applicable to human conditions (VandeBerg, 2004). Regarding research of ovarian cryopreservation for human fertility preservation, a similar tissue architecture and size of the ovary (Table 1) as well as being a mono-ovulatory species with the primordial follicles located superficially in the cortex would be of advantage (Gerritse et al., 2008). The previous research on ovarian cryopreservation using bovine, porcine, sheep and non-human primate models will be presented in this chapter.

Species	Ovarian volume	Tissue architecture	Ovulation pattern	Cycle length (days)
Cow	14.3 (+/- 5.7) cm³ (Gerritse et al. 2008)	Similar to the human	Mono, di-ovulatory	21
Pig	7.3 (+/- 2.2) cm³ (Gerritse et al. 2008)	Less fibrous than the human ovary	Multi ovulatory	18-24
Sheep	1.0 (+/- 0.4) cm³ (Gerritse et al. 2008)	Similar to the human	Mono, tri-ovulatory	16-17 (Seasonal)
Non-human primate (cynomolgus macaque)	0.258 (+/- 0.159) cm³ (Jones, 2011)	Similar to the human	Mono-ovulatory	28-32

Table 1. Ovarian characteristics of different large animal models

2. Cow

Cow ovaries may be used in research on ovarian cryopreservation. They are much larger (14.3+/- 5.7 cm³) (Gerritse et al., 2008) as compared to the human, but on the other hand the tissue architecture is similar. The function (Yang&Fortune, 2006) and the structure (Rodgers&Irving-Rodgers, 2010) of the bovine ovary has been extensively studied. Also, the granulosa cells and the extracellular matrix of bovine follicles of various developmental stages are well described (Lavranos et al., 1994; Irving-Rodgers et al., 2006).

There exist one report of avascular ovarian transplantation in the cow after cryopreservation by vitrification using a solution consisting of 20% ethylene glycol (EG) and 20% dimethylsulphoxide (DMSO) in TCM-199 medium (Kagawa et al., 2009). In this study, vitrification was performed using the Cryotissue method (Kagawa et al., 2007), where the ovarian cortex was positioned on a thin metal strip that was plunged into liquid nitrogen (LN) and inserted into a protective container for storage in LN. Rapid post-thaw warming was done with immersion of the metal strip into TCM-199, supplemented with sucrose at 40°C followed by washing in the identical solution but with decreasing sucrose concentration. After warming, the tissue was grafted subcutaneously to the neck or orthotopically to oophorectomized cows, and resumed cyclicity was seen in both groups within two months. Histological analysis of grafted tissue showed normal morphological appearance and about 80% viability among preantral follicles, as demonstrated by fluorescent staining. These results may be regarded as encouraging towards clinical application of the vitrification procedure for ovarian cortex cryopreservation.

In another study, aimed towards in vitro follicle maturation of bovine follicles, slow-frozen bovine ovarian cortical pieces were incubated after thawing for durations between 1 and 48 h (Paynter et al., 1999). The major finding was that the thawed tissue had a capacity to recover from damage during the subsequent incubation period. This idea was further utilized in later research, where bovine ovarian cortical pieces were cultured for six days followed by isolation of secondary follicles and culture in the presence of inhibin (McLaughlin&Telfer, 2010) showing significant estradiol (E2) secretion and oocytes growth up to a diameter > 100 μm.

The notion that antioxidants may enhance survival of frozen-thawed tissue was studied using bovine ovarian cortex (Kim et al., 2004). After cryopreservation by slow freezing with 1.5 M DMSO, bovine ovarian cortex was in vitro cultured for periods up to 48 h in minimal essential medium (MEM) with or without ascorbic acid. Interestingly, there was no difference between the two groups in apoptosis rate evaluated by terminal deoxynucleotidyl transferase dUTP nick end labelling test (TUNEL) or deoxyribonucleic acid laddering. Nevertheless, protective effects by ascorbic acid were seen in stromal cells that were cultured for 24 h. In addition, this study also demonstrated that stromal cells are more susceptible to damage mechanisms than primordial follicles, which is a finding also observed in frozen-thawed human ovarian tissue (Gook et al., 1999; Hreinsson et al., 2003). The model of supplementation of antioxidant agents to the cryoprotectant (CPA) should be investigated further.

The toxic effect of various concentrations and types of CPAs that are frequently used for slow freezing was compared using bovine ovarian cortical strips (Lucci et al., 2004; Celestino et al., 2008). Among other CPAs, DMSO at 1.5 M and 3 M were evaluated in both studies.

While the study by Celestino and colleagues (Celestino et al., 2008) showed increased toxicity with rising concentration of DMSO, the other study (Lucci et al., 2004) showed slightly higher proportion of normal follicles in the 3 M DMSO group when assessed by conventional histology. However, ultrastructural analysis by transmission electron microscopy (TEM) revealed some irregularities in the cytoplasm of granulosa cells when 3 M DMSO was applied (Lucci et al., 2004).

There is one study on bovine ovarian cortical strips comparing slow freezing with vitrification and this study demonstrated higher efficiency of slow-freezing (Gandolfi et al., 2006). Furthermore, another study also demonstrated advantages of the slow-freezing method as compared to vitrification, when bovine ovarian cortical pieces were used, but on the other hand an advantage of the vitrification method was seen when whole ovaries with vasculature were used as the model system (Zhang et al., 2011). In the latter study, the effectiveness of the different cryo techniques was evaluated by Trypan blue test, histology as well as E2 and progesterone levels obtained from supernatant after in vitro culture of the tissue. The same research group (Zhang et al., 2011) performed controlled-rate slow freezing with DMSO of whole bovine ovary and compared different cooling rates and ice seeding temperatures. The cooling rate of 0.2°C/min and ice seeding temperature of -5°C showed superiority in comparison to different combinations of the cooling rates of 2°C/min and 0.1°C/min and the ice-seeding temperatures of -2°C, -5°C and -8°C.

Additionally, one study on bovine ovarian tissue was designed to evaluate the effect of the thickness of the ovarian cortex strip on follicular morphology after incubation for 20 min and slow freezing with 1.5 M propylene glycol (PROH) (Ferreira et al., 2010). Ovarian cortex pieces of 10 x 3 mm, with a thickness of either 2 or 4 mm, were compared and considerably higher proportion of normal follicles were found in the 2 mm group compared to the 4 mm in both fresh and cryopreserved tissue. This result may be explained by superior tissue impregnation with CPA in the 2 mm group, but the exact mechanisms remain to be clarified.

3. Pig

The pig is a species that has been used in biomedical research, particularly regarding development and training of surgical techniques for later use in the human. The reproductive cycle of the pig lasts for 18-24 days and generally 8-15 oocytes are released from each ovary at ovulation (Soede et al., 2011). The size of the pig ovary is about 7.3 (+/- 2.2) cm³ (Gerritse et al., 2008), which is comparable to the human ovarian size of 6.5 (+/- 2.9) cm³ (Munn et al., 1986). The equal ovarian size of the sow and human female, as well as the possibility to get fresh pig ovaries from slaughterhouses, renders the pig as a good model for ovarian cryopreservation research. Nevertheless, a fairly low number of studies in this area with the pig ovary as an experimental model have been performed, as shown below.

One recent study on pig ovarian tissue evaluated whether the size of ovarian cortical pieces is important for the cryopreservation outcome (Jeremias et al., 2003). Cortical strips were all of 1 mm thickness but either 1x1 mm or 5x1 mm in surface area. The pieces were cryopreserved by slow freezing in 1.5 M DMSO and after rapid thawing the size of the surviving primordial follicle pool, was compared to fresh tissue (1x1 mm) (Jeremias et al., 2003). The freezing method was uncontrolled-rate freezing with the cryovials containing ovarian tissue were placed in a freezer at -20°C for 30 min followed by plunging in LN vapor for 30 min and then stored in LN. The results of the experiment showed similar density of primordial follicles of the 5x1x1 mm group as compared to the fresh tissue, while lower number of primordial follicles was observed in the small (1x1 mm) frozen-thawed pieces. It was not further discussed why the larger pieces were more resistant to cryoinjury.

In one study, using porcine ovarian cortex samples, programmed slow-freezing was performed with four different CPAs (glycerol (GLY)-10%; DMSO-1.5 M, EG-1.5 M, PROH-1.5 M) (Borges et al., 2009). The ovarian cortex pieces were incubated after thawing, and histological analysis, including light microscopy and TEM, demonstrated that the follicular viability was decreased after freezing with better results obtained by DMSO and EG as compared to PROH and GLY. This result correlates to that pregnancies in the human so far only have been demonstrated after ovarian cryopreservation in either DMSO or EG as CPAs (Donnez et al., 2004; Andersen et al., 2008). However, it should be emphasized that species differences exist in this regard, as demonstrated in a comparative study between ovarian cortex of human, bovine and porcine ovarian tissue (Gandolfi et al., 2006). In that study, Gandolfi and colleagues study showed that DMSO and PROH were equally effective to protect primordial and primary follicles of the pig ovary to cryoinjury and the pig ovarian tissue was also more resistant to cryoinjury as compared to the bovine and the human ovarian tissue.

The pig model was also used to study vitrification procedures (Gandolfi et al., 2006; Moniruzzaman et al., 2009). In one study, ovarian strips from 15-day old pigs, were vitrified using 15% EG, 15% DMSO and 20% fetal calf serum with addition of either 0 M, 0.25 M or 0.5 M sucrose (Moniruzzaman et al., 2009). Histological evaluation after warming showed higher percentage of healthy primordial follicles when CPA solution with 0.25 M sucrose was applied as compared to 0 M and 0.5 M solution. Moreover, the higher oocyte shrinkage was observed with the sucrose supplementation of 0.5 M and these results can be explained by unsatisfactory cell dehydration without sucrose as well as to excessive dehydration at 0.5 M sucrose. In one study, xenografting of vitrified ovarian tissues under the kidney capsule of nude mice was performed (Moniruzzaman et al., 2009). Histological evaluation, two months after grafting, revealed decrease of the primordial follicle density by around 20% in both fresh and vitrified grafts, but in the vitrified grafts, the follicles did not developed beyond the secondary stage. Hence, it may be that follicular development is disturbed after vitrification or that a follicular stage dependant developmental blockage is incurred by cryoinjury. Another study using pig ovarian cortex demonstrated low survival rate of primordial follicles with either EG or a combination of EG and DMSO as CPA (Gandolfi et al., 2006)

The pig ovary has been used in one study of whole ovary cryopreservation (Imhof et al., 2004). The ovary was perfused through the ovarian artery with 1.5 M DMSO and cryopreserved with a slow freezing protocol. Light microscopy evaluation after thawing at 25°C demonstrated a lower proportion of viable primordial follicles in the frozen-thawed ovaries (84%) as compared to fresh controls (98%). Furthermore, about 20% of healthy primordial follicles were seen when the ovaries were positioned directly into LN without previously perfusion with CPA. Noteworthy is that TEM did not demonstrate any major cellular difference between the fresh and frozen-thawed tissue.

4. Sheep

A variety of in vivo and in vitro sheep models have been used in research on both ovarian cortex cryopreservation and whole ovary cryopreservation, most probably due to the large knowledge about the physiology of the sheep ovary. In addition, the sheep ovary has, similarly to the human ovary, a collagen-dense outer stroma containing the pool of primordial follicles (Arav et al., 2005). Nevertheless, the size of the sheep ovary is only around 20% of the human premenopausal ovary (Munn et al., 1986; Gerritse et al., 2008) and this fact has to be taken into consideration, particularly in comparative studies concerning whole ovary cryopreservation.

The pioneering study in ovarian cryopreservation was published in 1994 by the Edinburgh group (Gosden et al., 1994), where live births (Table 2), after ovarian cortex cryopreservation and transplantation in a large experimental animal, were reported for the first time. In fact, this study opened up the field for future clinical fertility preservation and this is also accentuated in its conclusion by stating "that frozen storage and replacement of a patient´s own ovarian tissue might be practicable when fertility potential is threatened by chemotherapy/radiotherapy". The cryopreservation protocol applied in that study has subsequently been widely used in research and clinical practice, often with minor modifications. It should be emphasized that this slow freezing-thawing protocol was adapted from a study on cryopreservation of mouse primordial follicles (Carroll&Gosden, 1993) and the authors also wrote that they "have no evidence whether it is optimal". In the Gosden study, ovarian cortex pieces were positioned in cryovials with Leibowitz L-15 solution containing donor serum and 1.5 M DMSO and the cryovials were held on ice for 15 min. The ovarian cortex strips were then considered equilibrated in the CPA and cooled in a programmable freezer at a rate of 2°C/min to -7°C and maintained at -7°C for 10 min before seeding. Freezing was performed further by reducing the temperature at a rate by 0.3°C/min to -40°C following by 10°C/min until -140°C before plunging into LN. Thawing was performed by exposure of frozen tissue to air temperature for 2 min and then placed in a water bath at room temperature for an unspecified time. The frozen-thawed slices were attached adjacent to the left ovarian pedicle at the second laparoscopy of the sheep. High progesterone level four months later indicated reestablished normal cyclicity. Eight months after grafting, one sheep gave birth and one more lamb was delivered by Cesarean section at gestational age of about 144 days.

In a follow up study the authors reported that the longevity of the transplants was at least 22 months, but at that time very few primordial follicles were found in the graft (Baird et al., 1999). The levels of follicle-stimulating hormone (FSH) and luteinizing hormone (LH) in animals which had received grafts were higher as compared to controls. However, all grafted animals demonstrated normal cyclical pattern. These findings were corroborated by the lower levels of inhibin, indicating reduced number of inhibin-producing small antral follicles. In the second part of the study Baird and colleagues grafted both fresh and frozen-thawed ovarian pieces under the kidney capsule of nude mice to assess whether the freezing-thawing procedure per se or if the warm ischemic time post transplantation was the main reason for loss of the follicle pool already within two years. A follicular depletion of about 65% was seen after grafting of fresh tissue as compared to fresh non-grafted ovarian tissue. Additionally 7% was lost after transplantation of frozen-thawed pieces. Thus, this study demonstrated that ischemia rather than cryopreservation is the main cause of follicle depletion.

Reference	Ovarian tissue	Cryopreservation	Cryoprotectant	Transplantation	Results
Gosden et al. 1994.	Cortical pieces	Slow freezing	1.5 M DMSO in Leibovitz L-15 with 10% donor calf serum	Avascular, orthotopic	2 live births out of 6 transplants
Salle et al. 2002, Salle et al. 2003.	Hemi-ovarian cortex	Slow freezing	2 M DMSO in BM1 medium with 10% fetal cord serum	Avascular, orthotopic	11 live births (of which 3 sheep delivered twins) out of 6 transplants; 2 sheep delivered for the second time
Bordes et al. 2005.	Hemi-ovarian cortex	Vitrification	2.62 M DMSO +2.6 M acetamide+1.3M propylene glycol+0.0075 M polyethylene glycol in BM1 medium	Avascular, orthotopic	4 live births (1 twins) out of 6 transplants
Imhof et al. 2006.	Whole ovary	Slow freezing	1.5 M DMSO in RPMI 1640 with 10% autologous sheep blood serum	Vascular, orthotopic	1 live birth out of 8 transplants

DMSO- dimethylsulphoxide; RPMI-Roswell Park Memorial Institute medium.

Table 2. Live-births after transplantation of cryopreserved ovarian tissue in large animal models

Depletion of the major follicular pool during the ischemic post grafting period was also demonstrated in a recent study in the sheep, where fresh and frozen ovarian cortical tissue were grafted either subcutaneously in the anterior abdominal wall or the uterine horn (Aubard et al., 1999). The cryopreservation procedure was a slight modification of the original Gosden method (Gosden et al., 1994). About 5% of primordial follicles survived grafting at evaluation after seven months. No pregnancy was demonstrated, but mature oocytes were retrieved after gonadotropin stimulation. In addition, poor fertilization rate and cleavage arrest at 4 cells were recorded in oocytes from both fresh and frozen ovarian tissue. These results open up the issue of about cytoplasmatic maturation as well as quality of the oocytes from grafted and frozen-thawed tissue.

In other studies, a larger part of sheep ovaries, containing also parts of the medulla, were evaluated regarding viability after cryopreservation and avascular transplantation (Salle et al., 1998). In that study, the ovarian tissue was cryopreserved using the slow freezing technique published by Gosden (Gosden et al., 1994). Transplantation was performed to the ovarian hilus. At evaluation after six months, well preserved morphology with follicles of all stages was demonstrated (Salle et al., 1998) and progesterone secretion was reestablished (Salle et al., 1999). The same research group reported live births (Table 2) after cryopreservation and avascular transplantation of hemi-ovarian cortex (Salle et al., 2002). However, the cryoprotocol had been slightly modified using higher concentration (2 M) of DMSO and the temperature was decreased by a rate of 2°C/min to -35°C followed by 25°C/min to -140°C. Semiautomatic seeding was initiated at -11°C. Orthotopic transplantation was performed. Four out of six ewes achieved pregnancy of which two delivered twins. The remaining four ewes were observed further for two years (Salle et al., 2003). All animals became pregnant (Table 2) and delivered lambs of which two sheep gave birth for a second time. Noteworthy is that considerable loss of follicles was seen in all grafts. The results of this study also emphasizes that the avascular transplantation per se induces major damage to the tissue.

In one study, vitrification of sheep hemi-ovaries was performed with the VS1 CPA solution, containing 2.62 M DMSO, 2.60 M acetamide, 1.31 M PROH and 0.0075 M polyethylene glycol in BM1 medium (Bordes et al., 2005). The ovarian tissue was equilibrated in increasing concentration of VS1 (12.5%, 25%, 50% and 100%) and then placed in cryovials containing 100% of VS1 followed by direct plunging into LN. The vitrified tissue was warmed in water at 37°C for 10 min. Notably, regained cyclicity was shown four months to one year after grafting and 3/6 sheep delivered offspring (Table 2). However, histological evaluation of the graft after delivery revealed very few follicles (6-58 follicles per graft).

A group from the Cleveland Clinic performed microvascular fresh ovary transplantation in the sheep (Jeremias et al., 2002) and soon after published their results after autotransplantation of the whole cryopreserved ovary applying the identical surgical technique (Bedaiwy et al., 2003) with comparison of the outcome to avascular grafting of frozen-thawed cortical pieces (frozen with Gosdens protocol). The whole ovaries were flushed by a solution consisting of Leibowitz L-15 medium, 10% fetal calf serum and 1.5 M DMSO and then allocated to controlled-rate freezing. The temperature was reduced at 2°C/min until seeding temperature of -7°C. Further reduction of temperature was done at 2°C /min until -35°C followed at by 25°C /min until -140°C when cryovials were positioned into LN. At thawing, the cryovials were placed in a water bath at 37°C. After that, the ovaries were flushed with Leibovitz L15 medium containing 10% fetal calf serum for 20 min followed by end-to-end anastomosis between the ovarian and the inferior epigastric vessels. Immediate vascular patency was showed in all grafts, but only 27% of the whole ovary transplants showed patent blood vessels eight to ten days after transplantation. Histological evaluation demonstrated large necrotic areas in the grafts of the non-patent anastomosis group and the primordial follicle pool was severely reduced in that group as compared to the patent group. Also, severe injuries with focal transmural necrosis of intraovarian vessels may be regarded as a cryoinjury of the vasculature. The follicular viability, evaluated by Trypan blue test, was around 80%, in the cortical avascular transplantation and the patent whole ovary transplantation group, with a slightly higher rate of apoptotic cells in the whole ovary group. It should be emphasized that this study evaluated only short-term (8-10 days) results.

The long-term outcome, using the same microsurgical technique as described above, was assessed in a study of ovarian viability approximately five months after transplantation of frozen-thawed ovaries (Grazul-Bilska et al., 2008). A similar cryoprotocol was applied as in the earlier experiments (Bedaiwy et al., 2003) and the only difference was that the temperature from the seeding point was decreased by the rate 0.3°C/min until -40°C in contrast to 2°C/min until -35°C in the previous study. Histological evaluation revealed normal follicular development in only 25% of the transplanted ovaries. Furthermore, oocytes (n=3) from the larger follicles could be matured, but fertilization was not achieved. The vessels of the patent grafts appeared normal, with expression of marker proteins such as factor VIII, vascular endothelial growth factor (VEGF) and smooth muscle cell actin (SMCA). In view of the fact that the normal follicular development was seen in only 25% of the transplanted ovaries, and that mature oocytes did not fertilize, the study points to that major improvements in the fields of whole ovary cryopreservation and retransplantation are needed.

Main advances in the whole ovary cryopreservation field may be to optimize the freezing technique and also to improve the anastomosis technique as evaluated by an Israeli research group (Revel et al., 2004). After perfusion via the ovarian artery with 1.4 M DMSO in University of Wisconsin (UW) solution for 3 min, the sheep ovary was cryopreserved by the directional freezing technology. This technique provides identical cooling rate through the whole organ and allows constant cooling rate. The temperature was decreased by a rate of 0.6°C/min until seeding and 0.3°C/min to the temperature of -30°C before placement in LN. Rapid thawing was accomplished by placement of the cryovials into a water bath at 68°C for 20 s and 37°C for 2 min. The ovary was transplanted by microvascular anastomosis to an orthotopic site by end-to-end anastomosis to the remaining parts of the ovarian vessels. Cyclicity was demonstrated in three out of eight animals at around six months after the procedure. At laparotomy eight weeks after surgery, adhesions were seen in only one animal. In a follow up study, presenting long-term outcome of this technique (Arav et al., 2005), three animals demonstrated cyclicity two to three years after transplantation. Oocytes obtained from these animals could be parthenogenically activated with divisions until the 8-cell stage. In an extensive long-term follow up study (6 years) of the three whole ovary transplanted sheep (Arav et al., 2010) two of the ovaries responded to gonadotropin stimulation and these ovaries were normal at post mortem histological evaluation. The third sheep did not respond to FSH stimulation and histology revealed a fibrotic ovary and absence of follicles. The importance of this study is that cryopreserved whole ovary can survive for a long time and indicates beneficial effect of directional slow-freezing method. Nevertheless, it should be pointed out that the ultimate end point of healthy offspring has not been demonstrated by the use of this cryopreservation technique.

There is only one report on live-birth after whole ovary cryopreservation and vascular transplantation in sheep (Imhof et al., 2006). In that study, the ovaries were cryopreserved using the protocol of Gosden and coworkers (Gosden et al., 1994), but naturally the ovaries were cannulated and perfused with CPA before controlled-rate slow-freezing. Thawing was done by exposure of the frozen ovaries in air for two minutes followed by placement in a water bath at 25°C for seven minutes and perfusion by Roswell Park Memorial Institute medium (RPMI) to remove CPA. Transplantation was performed after removal of the contralateral ovary and the frozen-thawed ovary was orthotopicaly transplanted using microvascular (9-0 sutures) anastomosis to the ovarian vessels which is a comparable technique to that described by the Israeli group (Arav et al., 2005). Cyclicity was reported in four out of nine transplanted sheep. Importantly, one pregnancy occurred in this resulted in delivery of a healthy lamb around 1.5 years after grafting. At histological examination of ovaries 18-19 months after transplantation, the size of the primordial follicle pool was only 2-8% of that in non transplanted control ovaries. The authors discussed that the major follicular loss occurred during the freezing-thawing procedure. However, this suggestion relied on results of histological assessment, which is probably an unreliable method to evaluate viability and should be combined with other methods. Nevertheless, this single large animal species live-birth after whole ovary cryopreservation is a proof of the concept, which should encourage further research in this area.

It is of considerable importance to recognize the mechanisms behind the low success rate of whole ovary cryopreservation and also to understand what cell compartments are affected by the cryopreservation and thawing procedures. It seems that the follicular survival and the ovarian function are directly correlated to the vascular patency, as demonstrated in one elegant study of heterotopic autotransplantation of the frozen-thawed whole ovary (Onions et al., 2009). The CPA and freezing protocols were similar to the traditional Gosden protocol (Gosden et al., 1994) and the microvascular anastomosis was by aortic patch and utero-ovarian vein to carotid artery and jugular vein, as developed more than 40 years ago (Goding et al., 1967). The control group was the animals which received non-frozen heterotopic transplants. Eight months after transplantation, 7/8 cryopreserved transplants and 3/4 fresh ovarian transplants demonstrated patency. However, regardless of vascular patency, 5/7 frozen-thawed ovaries with vascular patency did not regain cyclicity and eight months after transplantation, a follicular loss of about 90% was seen in both the fresh and frozen group. A possible damage of endothelial cells during cannulation was discussed as one detrimental factor.

In vitro studies of vitrification of the whole sheep ovary preceded the trials in vivo. It was demonstrated in an elaborate study that the whole sheep ovary could be vitrified and that the VS4 cryoprotectant solution (mixture of 2.75 M DMSO, 2.76 M formamide and 1.97 M PROH) was superior to the VS1 solution (mixture of 2.62 M DMSO, 2.60 M acetamide, 1.31 M PROH and 0.0075 M polyethylene-glycol) (Courbiere et al., 2005). After thawing, the higher primordial follicular density (50% vs 23%) as well as proportion of histologically normal primordial follicles (53% vs 25%) was demonstrated in theVS4 group. However, it should be underlined that endothelial damage of the vascular pedicle was more evident in the VS4 group.

In a subsequent study by the same research group, the thermodynamic properties of VS4 in RPS-1 medium were studied (Courbiere et al., 2006). Evaluation of the cooling rate was done by differential scanning calorimeter by connection of the thermocouples to the ovarian cortex, the medulla and the CPA solution. The rate of cooling was above 300°C/min, with the measured cooling rate of the cortex being slightly higher than that of the medulla. Furthermore, the cooling rate of the CPA solution was higher in comparison to the ovarian medulla and cortex. This finding may be explained by differences in tissue architecture and vascularity of the ovary that leads to uneven distribution of the CPA. Results of this study also pointed out that it is not likely that the ovarian tissue can be completely vitrified at the end of the procedure. Ice crystallization during warming was also observed, indicating that the warming rate did not exceed the critical warming rate. Nevertheless, in contrast to the former study (Courbiere et al., 2005), injury of the endothelial layer of the ovarian vasculature was not demonstrated which possibly may be the result of the two-step-warming procedure to avoid ice crystallization (Pegg et al., 1997) as used in this study (Courbiere et al., 2006).

The same research group demonstrated in a subsequent study that the warming rate of the cortex was slightly higher than the medulla (Baudot et al., 2007). This corroborates the irregular distribution of CPA as well as complexity of the ovarian tissue. As in the previous study (Courbiere et al., 2005), a primordial follicle survival rate of about 50% was demonstrated after warming. In addition, ice crystals were observed during the cooling and the authors discussed that maybe limited ice crystallization can be acceptable for clinical application of this procedure.

Subsequent to the in vitro research on whole ovary vitrification, as described above (Courbiere et al., 2005; Courbiere et al., 2006), the efficiency of this method in vivo was assessed. In one study only one out of five vitrified-warmed ovaries resumed endocrine function (Courbiere et al., 2009) after orthotopic vascular transplantation. Vitrification was obtained using VS4 solution (2.75 M DMSO, 2.76 M formamide and 1.97 M PROH) in RPS-1 medium and the ovaries were perfused by gradually increasing concentrations of CPA. At warming, the ovaries were kept in LN vapor following by placement in a water bath at 45°C. The rationale behind this two-step warming was to attempt to avoid fractures of the vessels during the glassy state (Pegg et al., 1997). A total follicular loss in the vitrified group was demonstrated one year after transplantation. Vascular thrombosis occurred in three out of four vitrified ovaries, with patent vessels seen in the fresh group. One possible explanation may be that the warm ischemic time was longer in the vitrified group (median 287 min) in comparison to the control group (median 129 min) although it is more likely that major injuries occurred during the cryopreservation procedures.

It is obvious from the results presented above that there is a need for more systematic studies on the different stages of cryopreservation/transplantation procedures to get better results. As it relates to whole ovary cryopreservation, the viability and ovarian function should be evaluated after each stage of the procedure.

The effect of diverse cryoprotocols with accent of CPA toxicity was studied using the sheep hemiovary model (Demirci et al., 2001). The hemiovaries were incubated for 10 minutes in various concentrations (2-10 M) of DMSO and PROH and evaluated before and after slow-freezing-thawing (1, 1.5 and 2 M CPAs) concerning primordial follicle survival (Trypan blue test and histology). The follicular survival after incubation was higher at decreasing concentration of the CPA regardless of the type of CPA.

In vitro ovarian perfusion methodology, which was initially developed for evaluation of ovarian physiology (Brannstrom et al., 1987), may be used for evaluation of frozen-thawed ovaries (Fig. 1). The in vitro perfusion system highly mimics the physiological in vivo situation, and has shown that a complex processes, such as ovulation, occur during in vitro perfusion (Lofman et al., 1989). Recently, in vitro ovarian perfusion was used together with live-dead assay, histology and cell culture to assess the viability of frozen-thawed whole sheep ovaries (Wallin et al., 2009). The ovaries were frozen by slow uncontrolled-rate freezing (Martinez-Madrid et al., 2004) using PROH, stored in LN and then thawed in a water bath at 37°C. The in vitro perfusion results demonstrated compromised ovarian function of the ovary after cryopreservation in PROH, when compared to fresh controls.

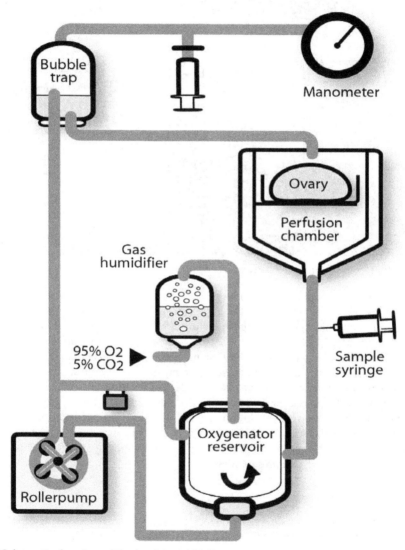

Fig. 1. Schematic drawing of the in vitro perfusion system

In another study of the same research group, the sheep ovaries were cryopreserved using the same uncontrolled-rate slow freezing with DMSO (Milenkovic et al., 2011). The ovaries in control group were frozen without CPA. Interestingly, steroid production after in vitro perfusion (Fig. 2) and cell culture was also demonstrated in control group which can be explained by adequate equilibration between cell dehydration and extracellular ice formation. However, a clear benefit of DMSO presence was seen.

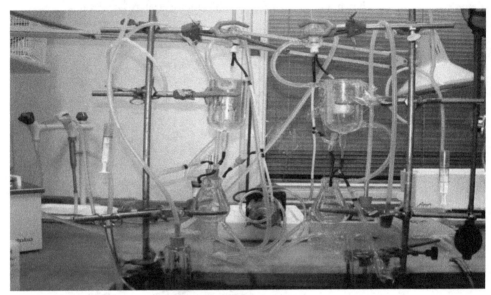

Fig. 2. Photograph of the in vitro perfusion apparatus

As discussed above, cryopreservation injury may not only occur within the follicles but also in the vascular bed of the ovary or in the large vessels. It was hypothesized that perfusion with the anti-apoptopic agent sphingosine-1-phosphate (S-1-P) before cryopreservation-perfusion by 1.5 M DMSO (Onions et al., 2008) may protect the tissue and diminish the cryoinjury. Histological examination after thawing showed arterial endothelial disruption of the vascular pedicle tissue in the cryopreserved ovaries, with the most extensive injury in the area where the cannula had been placed, followed by the hilus region and with less extensive injury on the venous side. No protective effect of the addition of sphingosine-1-phosphate could be demonstrated. In summary, this study was able to demonstrate the vulnerable state of the vasculature in whole ovary cryopreservation especially on the arterial side. It was also shown by proliferation and apoptosis markers that granulosa cells of antral follicles remain viable after cryopreservation.

5. Non-human primates

A number of non-human primate species have been used in research involving reproductive physiology and development of methods later used clinically in reproductive medicine. The benefit of these experimental animals is that the physiology and anatomy are similar to the human (Stevens, 1997; Weinbauer et al., 2008), although the ovarian size of the most studied cynomolgus monkey is considerably smaller ($0.258 +/- 0.159$ cm^3) (Jones et al., 2010) as compared to the human. Even if a lot of procedures are introduced in the human without appropriate tests in non-human primate species, it is advisable to include these animal models in preclinical research.

There is one study on non-human primate that has assessed the function of cryopreserved ovarian cortex tissue in vivo. The ovarian pieces were cryopreserved by controlled-rate freezing with 1.5 M DMSO and both fresh and frozen-thawed strips were grafted subcutaneously to the upper arm of cynomolgus macaque monkeys (Schnorr et al., 2002). Regained menstrual cycle and ovarian steroidogenesis were demonstrated in 80% of the animals after fresh ovarian cortex transplantation and in 50% of the animals after cryopreserved transplants, demonstrating that the cryopreservation technique may have damaging effect on the tissue. Additionally, only one mature oocyte was aspirated from subcutaneous ovarian tissue after gonadotropin stimulation performed in four animals. In the other part of this study (Schnorr et al., 2002) an attempt to enhance angiogenesis was performed by local administration of VEGF for two weeks after grafting, but no beneficial effect could be demonstrated.

There is one study on ovarian cortex from macaques when vitrification (3.4 M GLY, 4.5 M EG) and slow freezing (1.5 M EG) were compared and evaluated whether post-thawing co-culture on feeder cells (mouse fetal fibroblast monolayer) with addition of follicle stimulating hormone, insulin, transferrin and selenium, would increase viability (Yeoman et al., 2005). The post-thaw viability, as assessed by live-dead fluorescent staining, was comparable (around 70%) in the two groups, which was only marginally lower than the follicular viability of the fresh tissue (76%). Interestingly, follicular viability was increased after post-thawing co-culture, indicating rescue of partly damaged follicles and possible beneficial effect of co-culture.

In another study on ovarian cortex tissue of rhesus macaques, controlled-rate slow-freezing with 1.5 M EG was done and expressions of activin subunits as well as the phosphorylated form of the signalling protein were investigated (Jin et al., 2010). Activin subunits and the phosphorylated form of the signalling protein are markers for early follicular development. Immunohistochemistry revealed that these proteins were regularly distributed in primordial and primary follicles of both cryopreserved and fresh cortex. One interesting finding was that a higher rate of post thawing apoptosis was found in the stromal cells as compared to oocytes and granulosa cells. Another recently published study was able to demonstrate follicular development after in vitro three dimensional culture of cryopreserved secondary follicles from rhesus macaque (Ting et al., 2011). The ovarian cortex from rhesus macaque was cryopreserved by slow freezing (1.5 M EG) and vitrification (3 M GLY and 4.5 M EG). After thawing, the secondary follicles were mechanically isolated, encapsulated in 0.25% alginate and cultured for five weeks in MEM at 37°C. The development of antral follicles, although delayed and functionally compromised as compared to fresh tissue was demonstrated after both cryopreservation methods with slightly better results obtained after vitrification in comparison to slow freezing protocol. This observation again supports the need for further improvement of cryopreservation techniques.

6. Conclusions

Regardless pioneering success and reporting pregnancies after cryopreservation and avascular transplantation of ovarian tissue in human (Donnez et al., 2011) the future course regarding cryopreservation of ovarian tissue should be focused on further research using animal models. The size of ovarian cortex pieces, whole ovary freezing or not, choice and exposure time to CPA, stepwise or direct addition of CPA are some of the unresolved questions. The difference between species, complexity of ovarian tissue and uneven distribution of primordial follicles should be taken into consideration when creating an optimal study design. Vitrification seems promising, but differences between the protocols used make the interpretation of the data difficult. However, further research on animal models should lead to better understanding and improvement of cryopreservation techniques and then to higher efficiency, when used as a clinical procedure.

7. References

Andersen, C. Y., M. Rosendahl, A. G. Byskov, A. Loft, C. Ottosen, M. Dueholm, K. L. Schmidt, A. N. Andersen&E. Ernst. (2008). Two successful pregnancies following autotransplantation of frozen/thawed ovarian tissue. *Hum Reprod*, 23, 2266-2272, 1460-2350 (Electronic)

Arav, A., Z. Gavish, A. Elami, Y. Natan, A. Revel, S. Silber, R. G. Gosden&P. Patrizio. (2010). Ovarian function 6 years after cryopreservation and transplantation of whole sheep ovaries. *Reprod Biomed Online*, 20, 48-52, 1472-6491 (Electronic) 1472-6483 (Linking)

Arav, A., A. Revel, Y. Nathan, A. Bor, H. Gacitua, S. Yavin, Z. Gavish, M. Uri&A. Elami. (2005). Oocyte recovery, embryo development and ovarian function after cryopreservation and transplantation of whole sheep ovary. *Hum Reprod*, 20, 3554-3559, 0268-1161 (Print)

Aubard, Y., P. Piver, Y. Cogni, V. Fermeaux, N. Poulin&M. A. Driancourt. (1999). Orthotopic and heterotopic autografts of frozen-thawed ovarian cortex in sheep. *Hum Reprod*, 14, 2149-2154, 0268-1161 (Print) 0268-1161 (Linking)

Baird, D. T., R. Webb, B. K. Campbell, L. M. Harkness&R. G. Gosden. (1999). Long-term ovarian function in sheep after ovariectomy and transplantation of autografts stored at -196 C. *Endocrinology*, 140, 462-471, 0013-7227 (Print)

Baudot, A., B. Courbiere, V. Odagescu, B. Salle, C. Mazoyer, J. Massardier&J. Lornage. (2007). Towards whole sheep ovary cryopreservation. *Cryobiology*, 55, 236-248, 1090-2392 (Electronic)

Bedaiwy, M. A., E. Jeremias, R. Gurunluoglu, M. R. Hussein, M. Siemianow, C. Biscotti&T. Falcone. (2003). Restoration of ovarian function after autotransplantation of intact frozen-thawed sheep ovaries with microvascular anastomosis. *Fertil Steril*, 79, 594-602, 0015-0282 (Print)

Bordes, A., J. Lornage, B. Demirci, M. Franck, B. Courbiere, J. F. Guerin&B. Salle. (2005). Normal gestations and live births after orthotopic autograft of vitrified-warmed hemi-ovaries into ewes. *Hum Reprod*, 20, 2745-2748, 0268-1161 (Print)

Borges, E. N., R. C. Silva, D. O. Futino, C. M. Rocha-Junior, C. A. Amorim, S. N. Bao&C. M. Lucci. (2009). Cryopreservation of swine ovarian tissue: effect of different cryoprotectants on the structural preservation of preantral follicle oocytes. *Cryobiology*, 59, 195-200, 1090-2392 (Electronic) 0011-2240 (Linking)

Brannstrom, M., B. M. Johansson, J. Sogn&P. O. Janson. (1987). Characterization of an in vitro perfused rat ovary model: ovulation rate, oocyte maturation, steroidogenesis and influence of PMSG priming. *Acta Physiol Scand*, 130, 107-114, 0001-6772 (Print)

Carroll, J.&R. G. Gosden. (1993). Transplantation of frozen-thawed mouse primordial follicles. *Hum Reprod*, 8, 1163-1167, 0268-1161 (Print) 0268-1161 (Linking)

Celestino, J. J., R. R. Dos Santos, C. A. Lopes, F. S. Martins, M. H. Matos, M. A. Melo, S. N. Bao, A. P. Rodrigues, J. R. Silva&J. R. De Figueiredo. (2008). Preservation of bovine preantral follicle viability and ultra-structure after cooling and freezing of ovarian tissue. *Anim Reprod Sci*, 108, 309-318, 1873-2232 (Electronic) 0378-4320 (Linking)

Courbiere, B., L. Caquant, C. Mazoyer, M. Franck, J. Lornage&B. Salle. (2009). Difficulties improving ovarian functional recovery by microvascular transplantation and whole ovary vitrification. *Fertil Steril*, 91, 2697-2706, 1556-5653 (Electronic) 0015-0282 (Linking)

Courbiere, B., J. Massardier, B. Salle, C. Mazoyer, J. F. Guerin&J. Lornage. (2005). Follicular viability and histological assessment after cryopreservation of whole sheep ovaries with vascular pedicle by vitrification. *Fertil Steril*, 84 Suppl 2, 1065-1071, 1556-5653 (Electronic)

Courbiere, B., V. Odagescu, A. Baudot, J. Massardier, C. Mazoyer, B. Salle&J. Lornage. (2006). Cryopreservation of the ovary by vitrification as an alternative to slow-cooling protocols. *Fertil Steril*, 86, 1243-1251, 1556-5653 (Electronic)

Demirci, B., J. Lornage, B. Salle, L. Frappart, M. Franck&J. F. Guerin. (2001). Follicular viability and morphology of sheep ovaries after exposure to cryoprotectant and cryopreservation with different freezing protocols. *Fertil Steril*, 75, 754-762, 0015-0282 (Print) 0015-0282 (Linking)

Donnez, J., M. M. Dolmans, D. Demylle, P. Jadoul, C. Pirard, J. Squifflet, B. Martinez-Madrid&A. Van Langendonckt. (2004). Livebirth after orthotopic transplantation of cryopreserved ovarian tissue. *Lancet*, 364, 1405-1410, 1474-547X (Electronic)

Donnez, J., S. Silber, C. Y. Andersen, I. Demeestere, P. Piver, D. Meirow, A. Pellicer&M. M. Dolmans. (2011). Children born after autotransplantation of cryopreserved ovarian tissue. a review of 13 live births. *Ann Med*, 43, 437-450, 1365-2060 (Electronic) 0785-3890 (Linking)

Dorsch, M., D. Wedekind, K. Kamino&H. J. Hedrich. (2004). Orthotopic transplantation of rat ovaries as a tool for strain rescue. *Lab Anim*, 38, 307-312, 0023-6772 (Print) 0023-6772 (Linking)

Ferreira, M., A. Bos-Mikich, N. Frantz, J. L. Rodrigues, A. L. Brunetto&G. Schwartsmann. (2010). The effects of sample size on the outcome of ovarian tissue cryopreservation. *Reprod Domest Anim*, 45, 99-102, 1439-0531 (Electronic) 0936-6768 (Linking)

Gandolfi, F., A. Paffoni, E. Papasso Brambilla, S. Bonetti, T. A. Brevini&G. Ragni. (2006). Efficiency of equilibrium cooling and vitrification procedures for the cryopreservation of ovarian tissue: comparative analysis between human and animal models. *Fertil Steril*, 85 Suppl 1, 1150-1156, 1556-5653 (Electronic)

Gerritse, R., C. C. Beerendonk, M. S. Tijink, A. Heetkamp, J. A. Kremer, D. D. Braat&J. R. Westphal. (2008). Optimal perfusion of an intact ovary as a prerequisite for successful ovarian cryopreservation. *Hum Reprod*, 23, 329-335, 1460-2350 (Electronic)

Goding, J. R., J. A. Mccracken&D. T. Baird. (1967). The study of ovarian function in the ewe by means of a vascular autotransplantation technique. *J Endocrinol*, 39, 37-52, 0022-0795 (Print)

Gook, D. A., D. H. Edgar&C. Stern. (1999). Effect of cooling rate and dehydration regimen on the histological appearance of human ovarian cortex following cryopreservation in 1, 2-propanediol. *Hum Reprod*, 14, 2061-2068, 0268-1161 (Print)

Gosden, R. G., D. T. Baird, J. C. Wade&R. Webb. (1994). Restoration of fertility to oophorectomized sheep by ovarian autografts stored at -196 degrees C. *Hum Reprod*, 9, 597-603, 0268-1161 (Print)

Grazul-Bilska, A. T., J. Banerjee, I. Yazici, E. Borowczyk, J. J. Bilski, R. K. Sharma, M. Siemionov&T. Falcone. (2008). Morphology and function of cryopreserved whole ovine ovaries after heterotopic autotransplantation. *Reprod Biol Endocrinol*, 6, 16, 1477-7827 (Electronic)

Hreinsson, J., P. Zhang, M. L. Swahn, K. Hultenby&O. Hovatta. (2003). Cryopreservation of follicles in human ovarian cortical tissue. Comparison of serum and human serum albumin in the cryoprotectant solutions. *Hum Reprod*, 18, 2420-2428, 0268-1161 (Print)

Imhof, M., H. Bergmeister, M. Lipovac, M. Rudas, G. Hofstetter&J. Huber. (2006). Orthotopic microvascular reanastomosis of whole cryopreserved ovine ovaries resulting in pregnancy and live birth. *Fertil Steril*, 85 Suppl 1, 1208-1215, 1556-5653 (Electronic)

Imhof, M., G. Hofstetter, H. Bergmeister, M. Rudas, R. Kain, M. Lipovac&J. Huber. (2004). Cryopreservation of a whole ovary as a strategy for restoring ovarian function. *J Assist Reprod Genet*, 21, 459-465, 1058-0468 (Print)

Irving-Rodgers, H. F., K. D. Catanzariti, W. J. Aspden, M. J. D'occhio&R. J. Rodgers. (2006). Remodeling of extracellular matrix at ovulation of the bovine ovarian follicle. *Mol Reprod Dev*, 73, 1292-1302, 1040-452X (Print) 1040-452X (Linking)

Jeremias, E., M. A. Bedaiwy, R. Gurunluoglu, C. V. Biscotti, M. Siemionow&T. Falcone. (2002). Heterotopic autotransplantation of the ovary with microvascular anastomosis: a novel surgical technique. *Fertil Steril*, 77, 1278-1282, 0015-0282 (Print)

Jeremias, E., M. A. Bedaiwy, D. Nelson, C. V. Biscotti&T. Falcone. (2003). Assessment of tissue injury in cryopreserved ovarian tissue. *Fertil Steril*, 79, 651-653, 0015-0282 (Print)

Jin, S., L. Lei, L. D. Shea, M. B. Zelinski, R. L. Stouffer&T. K. Woodruff. (2010). Markers of growth and development in primate primordial follicles are preserved after slow cryopreservation. *Fertil Steril*, 93, 2627-2632, 1556-5653 (Electronic) 0015-0282 (Linking)

Jones, J. C., S. E. Appt, S. R. Werre, J. C. Tan&J. R. Kaplan. (2010). Validation of multi-detector computed tomography as a non-invasive method for measuring ovarian volume in macaques (Macaca fascicularis). *Am J Primatol*, 72, 530-538, 1098-2345 (Electronic) 0275-2565 (Linking)

Kagawa, N., M. Kuwayama, K. Nakata, G. Vajta, S. Silber, N. Manabe&O. Kato. (2007). Production of the first offspring from oocytes derived from fresh and cryopreserved pre-antral follicles of adult mice. *Reprod Biomed Online*, 14, 693-699, 1472-6483 (Print) 1472-6483 (Linking)

Kagawa, N., S. Silber&M. Kuwayama. (2009). Successful vitrification of bovine and human ovarian tissue. *Reprod Biomed Online*, 18, 568-577, 1472-6491 (Electronic)

Kim, S. S., H. W. Yang, H. G. Kang, H. H. Lee, H. C. Lee, D. S. Ko&R. G. Gosden. (2004). Quantitative assessment of ischemic tissue damage in ovarian cortical tissue with or without antioxidant (ascorbic acid) treatment. *Fertil Steril*, 82, 679-685, 0015-0282 (Print)

Lavranos, T. C., H. F. Rodgers, I. Bertoncello&R. J. Rodgers. (1994). Anchorage-independent culture of bovine granulosa cells: the effects of basic fibroblast growth factor and dibutyryl cAMP on cell division and differentiation. *Exp Cell Res*, 211, 245-251, 0014-4827 (Print) 0014-4827 (Linking)

Lofman, C. O., M. Brannstrom, P. V. Holmes&P. O. Janson. (1989). Ovulation in the isolated perfused rat ovary as documented by intravital microscopy. *Steroids*, 54, 481-490, 0039-128X (Print) 0039-128X (Linking)

Lucci, C. M., M. A. Kacinskis, L. H. Lopes, R. Rumpf&S. N. Bao. (2004). Effect of different cryoprotectants on the structural preservation of follicles in frozen zebu bovine (Bos indicus) ovarian tissue. *Theriogenology*, 61, 1101-1114, 0093-691X (Print) 0093-691X (Linking)

Martinez-Madrid, B., M. M. Dolmans, A. Van Langendonckt, S. Defrere&J. Donnez. (2004). Freeze-thawing intact human ovary with its vascular pedicle with a passive cooling device. *Fertil Steril*, 82, 1390-1394, 0015-0282 (Print)

Mclaughlin, M.&E. E. Telfer. (2010). Oocyte development in bovine primordial follicles is promoted by activin and FSH within a two-step serum-free culture system. *Reproduction*, 139, 971-978, 1741-7899 (Electronic) 1470-1626 (Linking)

Milenkovic, M., A. Wallin, M. Ghahremani&M. Brannstrom. (2011). Whole sheep ovary cryopreservation: evaluation of a slow freezing protocol with dimethylsulphoxide. *J Assist Reprod Genet*, 28, 7-14, 1573-7330 (Electronic) 1058-0468 (Linking)

Moniruzzaman, M., R. M. Bao, H. Taketsuru&T. Miyano. (2009). Development of vitrified porcine primordial follicles in xenografts. *Theriogenology*, 72, 280-288, 1879-3231 (Electronic) 0093-691X (Linking)

Munn, C. S., L. C. Kiser, S. M. Wetzner&J. E. Baer. (1986). Ovary volume in young and premenopausal adults: US determination. Work in progress. *Radiology*, 159, 731-732, 0033-8419 (Print)

Onions, V. J., M. R. Mitchell, B. K. Campbell&R. Webb. (2008). Ovarian tissue viability following whole ovine ovary cryopreservation: assessing the effects of sphingosine-1-phosphate inclusion. *Hum Reprod*, 23, 606-618, 1460-2350 (Electronic)

Onions, V. J., R. Webb, A. S. Mcneilly&B. K. Campbell. (2009). Ovarian endocrine profile and long-term vascular patency following heterotopic autotransplantation of cryopreserved whole ovine ovaries. *Hum Reprod*, 24, 2845-2855, 1460-2350 (Electronic)

Paynter, S. J., A. Cooper, B. J. Fuller&R. W. Shaw. (1999). Cryopreservation of bovine ovarian tissue: structural normality of follicles after thawing and culture in vitro. *Cryobiology*, 38, 301-309, 0011-2240 (Print) 0011-2240 (Linking)

Pegg, D. E., M. C. Wusteman&S. Boylan. (1997). Fractures in cryopreserved elastic arteries. *Cryobiology*, 34, 183-192, 0011-2240 (Print) 0011-2240 (Linking)

Revel, A., A. Elami, A. Bor, S. Yavin, Y. Natan&A. Arav. (2004). Whole sheep ovary cryopreservation and transplantation. *Fertil Steril*, 82, 1714-1715, 0015-0282 (Print)

Rodgers, R. J.&H. F. Irving-Rodgers. (2010). Morphological classification of bovine ovarian follicles. *Reproduction*, 139, 309-318, 1741-7899 (Electronic) 1470-1626 (Linking)

Salle, B., B. Demirci, M. Franck, C. Berthollet&J. Lornage. (2003). Long-term follow-up of cryopreserved hemi-ovary autografts in ewes: pregnancies, births, and histologic assessment. *Fertil Steril*, 80, 172-177, 0015-0282 (Print) 0015-0282 (Linking)

Salle, B., B. Demirci, M. Franck, R. C. Rudigoz, J. F. Guerin&J. Lornage. (2002). Normal pregnancies and live births after autograft of frozen-thawed hemi-ovaries into ewes. *Fertil Steril*, 77, 403-408, 0015-0282 (Print)

Salle, B., J. Lornage, B. Demirci, F. Vaudoyer, M. T. Poirel, M. Franck, R. C. Rudigoz&J. F. Guerin. (1999). Restoration of ovarian steroid secretion and histologic assessment after freezing, thawing, and autograft of a hemi-ovary in sheep. *Fertil Steril*, 72, 366-370, 0015-0282 (Print)

Salle, B., J. Lornage, M. Franck, L. Isoard, R. C. Rudigoz&J. F. Guerin. (1998). Freezing, thawing, and autograft of ovarian fragments in sheep: preliminary experiments and histologic assessment. *Fertil Steril*, 70, 124-128, 0015-0282 (Print) 0015-0282 (Linking)

Santos, R. R., C. Amorim, S. Cecconi, M. Fassbender, M. Imhof, J. Lornage, M. Paris, V. Schoenfeldt&B. Martinez-Madrid. (2010). Cryopreservation of ovarian tissue: an emerging technology for female germline preservation of endangered species and breeds. *Anim Reprod Sci*, 122, 151-163, 1873-2232 (Electronic) 0378-4320 (Linking)

Schnorr, J., S. Oehninger, J. Toner, J. Hsiu, S. Lanzendorf, R. Williams&G. Hodgen. (2002). Functional studies of subcutaneous ovarian transplants in non-human primates: steroidogenesis, endometrial development, ovulation, menstrual patterns and gamete morphology. *Hum Reprod*, 17, 612-619, 0268-1161 (Print)

Soede, N. M., P. Langendijk&B. Kemp. (2011). Reproductive cycles in pigs. *Anim Reprod Sci*, 124, 251-258, 1873-2232 (Electronic) 0378-4320 (Linking)

Stevens, V. C. (1997). Some reproductive studies in the baboon. *Hum Reprod Update*, 3, 533-540, 1355-4786 (Print) 1355-4786 (Linking) Ting, A. Y., R. R. Yeoman, M. S. Lawson&M. B. Zelinski. (2011). In vitro development of secondary follicles from cryopreserved rhesus macaque ovarian tissue after slow-rate freeze or vitrification. *Hum Reprod*, 26, 2461-2472, 1460-2350 (Electronic) 0268-1161 (Linking) Vandeberg, J. L. (2004). Need for primate models in biomedical research. *Gynecol Obstet Invest*, 57, 6-8, 0378-7346 (Print) 0378-7346 (Linking) Wallin, A., M. Ghahremani, P. Dahm-Kahler&M. Brannstrom. (2009). Viability and function of the cryopreserved whole ovary: in vitro studies in the sheep. *Hum Reprod*, 24, 1684-1694, 1460-2350 (Electronic) 0268-1161 (Linking)

Weinbauer, G. F., M. Niehoff, M. Niehaus, S. Srivastav, A. Fuchs, E. Van Esch&J. M. Cline. (2008). Physiology and Endocrinology of the Ovarian Cycle in Macaques. *Toxicol Pathol*, 36, 7S-23S, 1533-1601 (Electronic) 0192-6233 (Linking)

Yang, M. Y.&J. E. Fortune. (2006). Testosterone stimulates the primary to secondary follicle transition in bovine follicles in vitro. *Biol Reprod*, 75, 924-932, 0006-3363 (Print) 0006-3363 (Linking)

Yeoman, R. R., D. P. Wolf&D. M. Lee. (2005). Coculture of monkey ovarian tissue increases survival after vitrification and slow-rate freezing. *Fertil Steril*, 83 Suppl 1, 1248-1254, 0015-0282 (Print)

Zhang, J. M., Y. Sheng, Y. Z. Cao, H. Y. Wang&Z. J. Chen. (2011). Cryopreservation of whole ovaries with vascular pedicles: vitrification or conventional freezing? *J Assist Reprod Genet*, 28, 445-452, 1573-7330 (Electronic) 1058-0468 (Linking)

Zhang, J. M., Y. Sheng, Y. Z. Cao, H. Y. Wang&Z. J. Chen. (2011). Effects of cooling rates and ice-seeding temperatures on the cryopreservation of whole ovaries. *J Assist Reprod Genet*, 28, 627-633, 1573-7330 (Electronic) 1058-0468 (Linking)

11

New Approaches of Ovarian Tissue Cryopreservation from Domestic Animal Species

Vanessa Neto[1] et al.[*]
Université de Lyon, VetAgro Sup – Veterinary Campus of Lyon, UPSP ICE
'Interactions between Cells and their Environment', Team Cryobio
France

1. Introduction

Cryopreservation has been attempted for most of the developmental stages of both male and female reproductive cells, ranging from the immature gametes residing in ovarian or testicular tissues through the mature oocytes and spermatozoa.

However, each variant of the reproductive cells has introduced their own problems, and it has been realized that many aspects of the particular physiology of the cells will dictate how they respond to cryopreservation. Both male and female gametes have acquired highly specialized structural components (essential to fertilization and development) that may respond to the freezing process in ways different from that of basic cell structures.

1.1 Cryopreservation of spermatozoa

Semen is one of the most practical means of storing germplasm due to its abundant availability and ease of application (Holt and Pickard, 1999; FAO, 2007). Stored spermatozoa could be introduced back into existing populations either immediately or decades or centuries afterwards. In this way, cryopreservation of spermatozoa associated with artificial insemination (AI) or *in vitro* fertilization (IVF) facilitates the management of domestic animals herds, especially cow dairy herds where it is now used since 60 years. Cryopreservation better allows the use of semen from genetically superior males of threatened livestock breeds and has the potential to protect existing diversity and maintain heterozygosity while minimizing the movement of living animals and the transmission of venereal diseases (Johnston and Lacy, 1995; Andrabi and Maxwell, 2007).

Spermatozoon is a very small cell containing low amounts of cytoplasm and consequently low quantity of water. Furthermore sperm nuclear material is compacted and protected

*Thierry Joly[1], Loris Commin[1], Pierre Bruyère[1], Anne Baudot[2],
Gérard Louis[2], Pierre Guérin[1] and Samuel Buff[1]*
[1]*Université de Lyon, VetAgro Sup – Veterinary Campus of Lyon, UPSP ICE*
'Interactions between Cells and their Environment', Team Cryobio, France
[2]*Université Paris Descartes, Sorbonne Paris Cité, UFR Biomedical, France*

against physical injuries. For these reasons, cryopreservation of spermatozoa gives excellent results in term of viability and fertility, and is today widely used in human and animal assisted reproduction.

Semen from most mammalian and a few avian species has been successfully frozen in the past several years (FAO, 2007). Indeed, much better results have been obtained with sperm cryopreservation than with oocytes (and embryos) in term of viability. For these reasons, sperm cryobanking is used since the middle of the last century in domestic animal species and lately in human. Numerous storage facilities such as the French National Cryobank were also created for the preservation of valuable or endangered species. However, protocols currently used to conserve semen are still suboptimal and cannot be easily applied across species (Woods et al., 2004). First-service conception rates vary drastically between different breeding programs, but on average conception rates are fairly high in cattle, pigs, goats, and sheep. In wild species cryopreservation of gametes is currently used to preserve endangered species or breeds, and to overcome some fertility troubles. Breed reconstruction solely from semen is possible through a series of back-cross generations; however, the entire genetics of the original breed will not be recovered (Boettcher et al., 2005).

In human, gametes cryopreservation has also been developed to overcome fertility troubles (eg. genital duct problems, sexual dysfunction ...). Sperm cryopreservation that has produced live birth has been available for over 50 years (Sherman and Bunge, 1953). Anonymous donor sperm banking has been a fundamental concept of reproductive medicine for several decades, and artificial insemination and donor sperm cryobanking are widely used in reproductive medicine centers (Critser, 1998). The availability of frozen donor sperm has become a mainstay for the treatment of serious male infertility worldwide. More recently, gametes cryopreservation has been used to preserve fertility of men (or women) submitted to gonadotoxic treatments and elective sperm cryopreservation programs have been provided from cancer patients all over the world. Using vitrification of sperm obtained from testicular biopsy, epididymal fluid, or a semen sample after electroejaculation could create new hope for infertile men (Edge et al., 2006).

1.2 Cryopreservation of testicular tissue

Cryopreservation of testicular tissue has been studied since about ten years in animals (Jezek et al., 2001; Jahnukainen et al., 2007; Milazzo et al., 2008; Zeng et al., 2009; Abrishami et al., 2010; Milazzo et al., 2010; Curaba et al., 2011) and human (Bahadur and Ralph, 1999; Bahadur et al., 2000; Guerin, 2005; Revel and Mejia, 2010). It is the only available solution for pre-pubertal boys who must receive a gonadotoxic treatment (eg. cancer therapy;(Keros et al., 2005; Keros et al., 2007; Wyns et al., 2007; Wyns et al., 2008; Wyns et al., 2010).

In contrast to the situation in the ovary, it is well established that spermatogonial stem cells exist in the testis and are responsible for maintaining spermatogenesis from puberty for the lifetime of the male. Human testicular cells might be harvested, cryopreserved before a gonadotoxic treatment and re-introduced into the testis upon its completion. Possibilities include transplantation back into the inactive testes (ipsigenic germ cell transplantation), maturation *in vivo* in another host (xenogenic germ cell transplantation), or *in vitro* spermatogenesis. Mature sperm could then be used for fertilization by ICSI.

Sperm obtained after stem cell transplantation were shown to be able to fertilize mouse oocytes. Fertile offspring were obtained through artificial reproductive technologies following the establishment of complete spermatogenesis by grafting testis tissue from newborn mice, pigs, or goats into mouse host (Honaramooz et al., 2002; Schlatt et al., 2002). Different freezing protocols have been developed in several species but without a clearly identified procedure (Travers et al., 2011). Testicular tissue from prepubertal boys facing gonadotoxic treatment could be banked for several years for spermatogonial stem cell transplantation. Pregnancies have been achieved with ICSI using immature spermatids and secondary spermatocytes extracted from testicular tissue in men with spermatogenic arrest (Fishel et al., 1997; Vanderzwalmen et al., 1997; Sofikitis et al., 1998).

1.3 Cryopreservation of oocytes

Oocyte cryopreservation offers many advantages. It permits to preserve endangered species and low effective breeds, and to preserve fertility of high genetic value females. In human it permits to overcome some fertility problems. Oocyte banks would also enlarge the gene pool, facilitate several assisted reproductive procedures, salvage female genetics after unexpected death, and avoid controversy surrounding the preservation of embryos (Ledda et al., 2001; Checura and Seidel, 2007). Like semen, oocyte cryopreservation is beneficial for international exchange of germplasm, as it avoids injury and sanitary risks involved in live animal transportation (Pereira and Marques, 2008). But oocyte cryopreservation gives much lower results when compared with spermatozoa.

This is due essentially to the much important size and complexity of oocytes. For example, nuclear material is much more exposed in prophase I or metaphase II oocyte than in compacted chromatin of sperm cells. Oocytes collected from slaughterhouse derived ovaries are at the germinal vesicle (GV) stage in which the genetic material is contained within the nucleus. Since this stage has no spindle present, GVs are assumed to be less prone to chromosomal and microtubular damage during cryopreservation. However, oocytes can also be cryopreserved at the metaphase II stage of maturation. During the metaphase II stage, the cumulus cells surrounding the oocyte are expanded, microfilaments of actin are involved in cell shape and movements, and microtubules form the spindle apparatus (Massip, 2003). Moreover, cryopreservation of oocytes necessitates the success of the following steps: in vitro maturation, in vitro fertilization and embryo culture. Progress in female gametes cryopreservation has gone hand in hand with that for in vitro maturation and embryo culture. In livestock animals, oocytes collected by in vivo pickup or at slaughter can be frozen for extended periods of time for subsequent IVF to produce embryos. However, in some species such as canidae the collection of oocytes is difficult and in vitro maturation (IVM), in vitro fertilization (IVF) and embryo culture are not yet under control (Luvoni et al., 2006). For these reasons precise cooling and thawing rates and the use of a programmable cell freezer are necessary for oocyte (and also embryo) cryoconservation and very few studies have been conducted in animals.

Oocytes are extremely sensitive to chilling, and the technique is not as established as in semen or embryos, due to the fact that oocytes typically have a low permeability to cryoprotectants (Woods et al., 2004). The major differences between oocytes and embryos are the plasma membrane, presence of cortical granules, and spindle formation at metaphase II stage of meiosis (Chen et al., 2006; Salvetti et al., 2010). To date, there has been no consistent oocyte cryopreservation method established in any species, although, there has been significant progress and offspring have been born from frozen-thawed oocytes in cattle, sheep, and horses (Otoi et al., 1996; Maclellan et al., 2002; Woods et al., 2004). During the process of cryopreservation, oocytes suffer considerable morphological and functional damage, although, the extent of cryoinjuries depends on the species and the origin (*in vivo* or *in vitro* produced). The mechanism for cryoinjuries is yet to be fully understood, and until more insight is gained, improvement of oocyte cryopreservation will be difficult. Also, it is noticeable that immature oocyte present in primordial follicles seems more resistant to cryopreservation when compared to mature oocyte. Consequently, cryopreservation of ovarian cortex may constitute an interesting alternative to isolated mature oocyte cryopreservation.

1.4 Cryopreservation of ovarian tissue

At birth, the ovaries contain the lifetime complement of primary oocytes which are arrested in the prophase stage of meiosis 1 and are surrounded by a single-layered epithelium to form the primordial follicles. Ovarian cortex presents several advantages when compared with isolated oocytes: (*i*) it contains the important pool of growing follicles; (*ii*) it does not necessitate the *in vitro* maturation /*in vitro* fertilization /embryo culture steps if it is associated with grafting; (*iii*) no previous ovarian stimulation is necessary. Consequently cryopreservation of ovarian cortex is an alternative to cryopreservation of isolated oocytes or embryos. It could be used as an emergency preservation and as infertility therapy method for valuable animals. Ovarian cortex cryopreservation has been developed in human in order to preserve fertility in young women submitted to gonadotoxic therapy (Stahler et al., 1976; Gook et al., 2004). In human newborns were obtained after orthotopic autograft of frozen-thawed ovarian cortices (Donnez et al., 2004).

It is obvious that, to achieve successful cryopreservation of ovarian tissue, it is essential to maintain the functional status of the whole mixture of different cell types: oocytes, granulosa cells, epithelial cells, fibroblasts... This represents a major difficulty, because the optimum kinetic of cooling is different for each cell type. Oocytes are large cells, with a low surface to volume ratio, surrounded by zona pellucida. Immediately adjacent to the oocyte are corona radiata cells that have long cytoplasmic extensions which penetrate the zona pellucida, ending in oocyte membrane. These processes and gap junctions are important in the metabolic cooperation between the oocyte and surrounding layers of granulosa cells, which form the cumulus-oocyte complex during the growth phase. Consequently, at the opposite to cryopreservation of isolated cells, a cryopreservation protocol for a tissue represents a compromise between the requirements of the different constitutive cells.

The early work on ovarian tissue cryopreservation was performed in animal studies: rabbit (Smith, 1952) and rat (Parkes and Smith, 1953; Deanesly, 1954). The earliest positive results were obtained when glycerol (15%) plus serum were used as cryoprotective agents (CPAs) for cryopreservation of rabbit granulosa cells, via a slow cooling protocol (Smith, 1952). An equilibration period was necessary to achieve CPA penetration into the tissue. For this reason small samples were recommended. A rapid rewarming by plunging the samples into a water bath at 40°C was the most effective procedure (Parkes and Smith, 1953; 1954). Normal offspring were obtained from mice with orthotopic ovarian grafts of tissue that had been frozen and stored at -79°C (Parrott, 1960). Vitrification of ovarian tissue was also investigated. Nevertheless, Isachenko et al suggested that in human, low freezing protocols were more promising than vitrification protocols (Isachenko et al., 2009).

This technique has also been developed in rabbit (Neto et al., 2007a), mouse (Candy et al., 1997), rat (Aubard et al., 1998), ewe (Gosden et al., 1994; Demirci et al., 2001), cow (Paynter et al., 1999). We have obtained newborn rabbits after autografting of cryopreserved ovarian cortex (Neto et al., 2007b). Also, our team developed this technique in cat (Neto et al., 2006) and dog (Commin et al., 2011).

Several techniques have been applied to ovarian cortex cryopreservation: slow freezing, vitrification. Simultaneously to ovarian tissue cryopreservation, numerous researches have been conducted about ovarian tissue grafting: orthotopic, heterotopic, auto-, allo- and hetero-grafting (Pullium et al., 2008).

2. Development of optimized methods for the cryopreservation of the ovarian tissue in domestic mammalian species

The most common cryopreservation method is the slow freezing procedure, consisting of an initial slow, controlled-rate cooling to subzero temperatures followed by rapid cooling as the sample is plunged into liquid nitrogen for storage (−196°C). At such a low temperature, biological activity is effectively stopped, and the cells functional status may be preserved for centuries. However, several physical stresses damage the cells at these low temperatures. Intracellular ice formation is one the largest contributors to cell death; therefore, freezing protocols use a combination of dehydration, freezing point depression, supercooling, and intracellular vitrification in an attempt to avoid cell damage.

Currently used ovarian cortex cryopreservation protocols have been direct, or slight modifications of the methods developed for isolated oocytes and embryos. There were primarily developed by trial and error adjustments of cooling and warming rates, and choice of CPA and CPA concentrations. However, because there are a large number of protocol variables potentially affecting cell viability, an exhaustive experimental search for the optimal combination of these parameters has long been considered to be prohibitively expensive in terms of time and resources.

2.1 Chemical and physical parameters affecting equilibration and freezing processes of ovarian tissue in mammalian species

The result of a cryopreservation process is influenced by several chemophysical parameters affecting directly or not the functions and the integrity of the ovarian cells along the freezing process, from the equilibration to the thawing. Among these parameters, the method of equilibration, the freezing rate, the composition of the freezing solution and notably the nature of the permeating CPAs and the non-permeating CPAs, the concentration of each CPA, the use of serum, or the rate of thawing may be investigated to know the relative influence of each of them and the induced cell injuries.

In general, we can expect coupled flows of water and CPAs when CPAs are added, during freezing, thawing and when CPAs are removed from the cells, resulting in a series of anisosmotic conditions. During freezing, the cells dehydrate and shrink and remain shrunken during storage, but return to their isosmotic volume upon thawing. Finally, the cells are subjected to potentially lethal swelling upon CPA dilution and removal. During the controlled slow cooling extracellular ice formation is induced (seeding) at a temperature just below the solutions' freezing point, and then the cooling continues at a given rate in the presence of a growing extracellular ice phase, which raises the extracellular solute concentration in the unfrozen fraction and results in water being removed from the cell via exosmosis.

Permeating CPAs, such as glycerol, dimethyl sulfoxide, ethylene glycol or propylene glycol are typically included in the cryoprotective medium, to protect the cells against injury from the high concentrations of electrolytes that develop as water is removed from the solution as ice. During the equilibration step the inner cell water is partly replaced by the permeating CPAs. However, the CPAs can be damaging to the cells, especially when it is used at high concentrations. The toxicity can be reduced by decreasing the time or the temperature of the equilibration step (Karlsson and Toner, 1996). But equilibration at low temperatures requires increasing the exposition time to freezing solution. Furthermore, the CPAs may have dramatic osmotic effects upon the cells during their addition and their removal. Consequently, the use of several steps of increasing concentrations of CPAs during the equilibration allows reducing the osmotic gradient. The cells exposed to such permeating CPAs undergo initial dehydration, followed by rehydration, and potential gross swelling upon removal. This osmotic shock may generate membrane damages by mechanical means and predisposition of the cell to injuries during the other steps of cryopreservation, or even cell death (Mazur and Schneider, 1986). These kinds of damages could be reduced by using cells surfactant such as serum. During the freezing step, the follicular preservation depends on the nature and the concentration of the CPAs.

Control of the cooling and warming rates is also crucial, as the freezing/thawing rates and the temperature of seeding also influence the ice properties. If cells are cooled too rapidly during the controlled slow cooling process, water does not exit the cells fast enough to maintain equilibrium and, therefore, the oocytes and other ovarian cells freeze intracellularly, resulting in death in most cases. If cooling is too slow, the long duration can cause 'solution effects' injury resulting from the high concentration of extra- and intracellular solutes, probably due to the effects of the solutes on the cellular membrane or through osmotic dehydration. During warming the small intracellular ice crystals might subsequently undergo recrystallization, forming bigger ice crystals that rupture the cell membrane, thus leading to fatal damage. Finally, the thawing and the removal of the CPA depend on the temperature and on the presence of non-permeating CPA limiting the osmotic swelling during rinsing.

2.2 Use of fractional experimental design

The influence of the multiple chemical and physical parameters cannot be exhaustively performed as it would require too much time and resources. Even if the number of factors, k, in a design is small, the 2^k runs specified for a full factorial can quickly become very large. For example, $2^5 = 32$ runs is for a two-level, full factorial design with five factors. To this design we would need to add a good number of centerpoint runs and we could thus quickly run up a very large resource requirement for runs with only a modest number of factors.

Moreover, while the approach is sequential in nature, it is potentially increasing in complexity as the knowledge and understanding of the application and domain evolves. Design of experiments techniques provides a systematic, effective and efficient approach to the investigation of a phenomenon. The main advantages of this strategy were the saving in times and resources expended compared to other approaches and the resulting mathematical models that helps us to better understand the phenomena under investigation more fully. To analyze the response of ovarian cortices from different species to the freeze/thaw process, the authors decided to use fractional (2^{n-p}) experimental designs (Mechakra et al., 1999).

Using this statistical tool, the authors used only a fraction of the runs specified by the full factorial design; which runs to make and which to leave out was one of our subjects of picked. The authors used various strategies that ensure an appropriate choice of the runs. As for an example, fractional experimental designs 2 [(5-2)] presented in table 1 aim to evaluate the combined effect of five different factors according to two modalities for each of them. For each experimental design, eight combinations of factors were performed. For each of them, the ratio of morphological preservation and the ratio of viability of isolated preantral follicles were recorded. While the designs were similar for each of the species that were evaluate, the parameters were chosen according to each species (Table 2 to Table 5).

Run	Variable						
	I	X_1	X_2	X_3	X_4	X_5	$X_1.X_2$
1	+1	-1	-1	-1	+1	-1	+1
2	+1	+1	-1	-1	+1	+1	-1
3	+1	-1	+1	-1	-1	+1	-1
4	+1	+1	+1	-1	-1	-1	+1
5	+1	-1	-1	+1	-1	+1	+1
6	+1	+1	-1	+1	-1	-1	-1
7	+1	-1	+1	+1	+1	-1	-1
8	+1	+1	+1	+1	+1	+1	+1

Table 1. Fractional experimental design $2^{(5-2)}$ used to evaluate the cryopreservation protocols in the doe rabbit, in the queen and in the cow

Variables	Level -1	Level +1
X_1: Permeating CPA	DMSO	PROH
X_2: Concentration of permeating CPA	1.5M	2M
X_3: Non permeating CPA	trehalose	sucrose
X_4: Freezing rate	0.3°C/min	2°C/min
X_5: Equilibration	1 step	3 steps

Table 2. Dependent variable list evaluated in the rabbit doe

These $2^{(n-2)}$ experimental designs allowed discriminating between five factors influencing the cryopreservation process (variables X_1 to X_5) and the simultaneous interactions between two of them. The linearity (structural and estimated model) of the experimental model was evaluated by an ANOVA test. One randomly chosen assay was replicated three times to estimate the experimental error (E).

So, eleven experiments were randomly performed. Multi-linear regression was performed using all the variables in order to evaluate experimental results according to this model:

$$\hat{y} = \beta_0 + \beta_1.X_1 + \beta_2.X_2 + \beta_3.X_3 + \beta_4.X_4 + \beta_5.X_5 + \beta_{1,2}.X_1.X_2 + E$$

Variables	Level -1	Level +1
X1: Permeating CPA	PROH	DMSO
X2: add of sucrose	no	yes
X3: Freezing rate	0.5°C/min	2°C/min
X4: manual seeding	no	Yes
X5: Freezing rate after -40°C	In the freezing chamber	Direct immersion in LN$_2$

Table 3. Dependent variable list evaluated in the queen

Variables	Level -1	Level +1
X1: Permeating CPA	PROH	DMSO
X2: Concentration of permeating CPA	1.5M	2.5M
X3: Non permeating CPA	sucrose	trehalose
X4: Medium	Euroflush®	Medium with choline
X5: Cell surfactant	Fetal calf serum	Albumax®

Table 4. Dependent variable list evaluated in the cow

Variables	Level -1	Level +1
X1: Permeating CPA	DMSO	PROH
X2: Non permeating CPA	trehalose	sucrose
X3: Freezing rate	0.3°C/min	2°C/min
X4: Equilibration	1 step	3 steps

Table 5. Dependent variable list evaluated in the bitch

Results of experimental designs for each species were completed by at least one additional biological evaluation and one quantitative evaluation of normal and viable follicle rates after freezing, according to the best combination of factors chosen with experimental designs.

2.3 Criteria to assess the quality of frozen-thawed cortices

A survey of nearly all quality assays available to the preservation scientist reveals that they can be grouped into different categories. The following assay tier is not specific to cryopreservation, but is presented below as a support that can guide those who work with preserved tissues. The authors decided to assess the quality of the protocols developed in different species using all or part of alternate test presented below. The morphology and the viability of the ovarian follicles were systematically assessed, in combination with the investigation of the ultrastructure of the follicles, and their capacity to resume folliculogenesis after graft.

2.3.1 Morphology and ultrastructure

Assessment of the morphology of ovarian cells required the ovarian fragments were fixed in a preliminary defined fixative agent adapted to the species, before being processed for classical light microscopy. Primordial to primary follicles – from oocytes surrounded by flattened granulosa cells until oocytes surrounded by one layer of cuboidal granulosa cells (Gougeon and Chainy, 1987) – were usually classified into four types of morphological defects: *(Type I)* follicle without any morphological defect - follicle is regular, with joined follicular cells; cytoplasm of the oocyte is homogeneous and chromatin is diffused and regular; *(Type II)* follicle with cytoplasmic defect - cytoplasm of the oocyte is vacuolated or eosinophil; *(Type III)* follicles with nuclear defect - nucleus of the oocyte is picnotic, without apparent nuclear membrane or with an irregular nuclear membrane; *(Type IV)* degenerated follicle – oocyte with combined cytoplasmic and nuclear defects or follicle with irregular shape or with disjoined follicular cells or with swelled follicular cells.

The ultrastructure of ovarian follicles was also examined for the presence of apoptotic and para-apoptotic cell death. Ultrastructurally apoptosis is characterized by margination of condensed chromatin, nuclear fragmentation, and the formation of apoptotic bodies. Para-apoptosis, nonclassical apoptosis, is a specific morphologic type of non-necrotic cell death and is characterized by cytoplasmic vacuolization, condensed chromatin (but not early margination of the chromatin), and swollen mitochondria.

2.3.2 Cytolysis live/dead assay

The cytolysis assays have both a very positive and a negative attribute to them. On the positive side, there are a variety of assays that can reveal cell membrane leakage that occurs as a final stage in most forms of cell death. Yet given that cytolysis is the last stage of preservation-induced cell death, these assays do not reveal early-stage mechanisms underlying preservation-induced cell death and thus have limited use in the future as a diagnostic means to develop improved preservation formulations and protocols. The LDH assay continues today to be useful for measuring preservation-induced cytotoxicity. The concept behind this cytolysis assay is simply that if the cell membrane is compromised, then LDH will leak into the extracellular milieu where it can be measured. The trypan blue assay is also one of the most commonly used cytolysis assays. A number of investigators has used the trypan blue exclusion assay in studying preservation efficacy. It does however share the same handicap as the LDH assay given that neither can be analyzed using fluorescence and/or bioluminescence. Currently, the best cytolysis live/dead assays are those that employ fluorescent indicator dyes. Available probes can be subdivided into two different subsets, one of which is trapped by the cell and leaks out only is a membrane rupture occurs. The other subtype, exemplified by ethidium homodimer or propidium iodide, is membrane insoluble and only stains the cell if it gains access through a compromised plasma membrane. In this way, the authors did evaluate the viability of the frozen-thawed follicles using Calcein-AM and Ethidium homodimer I stains (Live/Dead® Viability/Cytotoxicity Kit, Molecular Probes) on enzymatically isolated follicles (Fig. 2).

2.3.3 In vivo functionality (autografting model) and in vivo growth potential (xenografting model)

One of the key components of any preservation protocol must be a functional assay that matches the type of cells or tissue being analyzed. In some cases this determination is quite easy. As for example, the sperm motility is a well-accepted functional assay for this system, whereas the capacity to resume folliculogenesis sounds reasonable for the ovarian cortex.

Fresh and cryopreserved rabbit doe ovarian tissues were autografted into young females. Fresh ovarian tissue was grafted on the ipsilateral ovarian pedicle immediately after ovary resection (2 grafts per female), contrary to cryopreserved ovarian tissue grafted on the controlateral ovarian pedicle 24 hours after freezing (one graft per female). Control females were ovariectomised, according to the same surgery resection that used before graft. From height week after graft, grafted rabbit doe were inseminated every 3 weeks in case of negative pregnant diagnosis. Eleven months later, ovarian grafts were removed at necropsy to observe follicular structures by histology.

In the bitch, the growth ability of frozen-thawed ovarian tissue has also been assessed by implantation of small pieces of ovarian tissue into adult female SCID mice. After removing mice ovaries, the canine frozen thawed ovarian tissue was placed intramuscularly in the gluteal superficial muscle. To study the setting up of graft resumption, the graft was harvested at one, eight or 16 weeks and the resumption of the ovarian activity was controlled by vaginal cytology assessment. After harvesting, the grafts were processed for histological assessment.

2.3.4 Vascularization

In the bitch, the alpha smooth muscle actin (alpha-SMA), a marker of the mature blood vessel, was used to assess the vessel density within the ovarian tissue, using a primary antibody directed against alpha-SMA. For each slice, image analysis was performed under direct light microscope to determine the tissue areas. The stained vessels were counted in several fields of the same slice and in several slices per animal to deduce a blood vessel density.

2.4 Experimental results

To illustrate the interest of the use of fractional experimental design, the authors decided to present some of their results obtained in the doe rabbit, in the queen, in the bitch and in the cow during the last 6 years. The doe rabbit was used as a model for the human and the animal applications of the ovarian tissue cryopreservation, because of its biological and breeding characteristics. The cat was considered as a model for the ovarian tissue cryopreservation studies of endangered wild felids; from all the felids, the domestic cat is the only non endangered feline species. The cow was used as model for ruminants, with a special interest for preserving high valuable individuals in combination to embryos and to semen cryopreservation. Finally, the bitch was studied for preserving the reproductive potential of future guide dogs submitted to neutering surgery before training.

2.4.1 In the doe rabbit

The experimental variability expressed as the repetition of one single combination, showed that both the structural and the estimated models of the experimental design were valid when considering the morphological preservation ratio of the follicles. The concentration of the permeating CPA ($P = 0.67$) and the number of equilibration steps ($P = 0.19$) seemed to have no significant effect on the morphological preservation ratio of ovarian follicles. The nature of the permeating and non-permeating CPA seemed to influence the morphological preservation ratio of the follicles ($P = 0.08$ and $P = 0.07$ respectively) although the non-significant difference. DMSO tended to reduce the morphological preservation ratio, as compared with PROH. Morphological preservation ratio was increased in the presence of trehalose compared with sucrose. The freezing rate seemed to be the factor that had the greatest impact on the morphological preservation ratio of the doe rabbit follicles. At a freezing rate of 0.3°C/min we observed a significant increasing of the follicular morphological preservation ratio, as compared with 2°C/min ($P<0.01$). No significant interaction was observed between the nature of the permeating CPA and its concentration.

According to the results of the experimental design, the precise evaluation of the best combination of factors influencing positively the morphological preservation ratio (3 steps equilibration protocol, 1.5M DMSO or 1.5M PROH, medium supplemented with either sucrose or trehalose) was performed. Ovarian pieces were treated according to the results obtained with experimental design. Ovarian cortices were equilibrated (3 steps) in the freezing media based on TCM 199 and 10% FCS, at room temperature. The freezing media was supplemented with 1.5 M DMSO or 1.5M PROH and 0.2 M sucrose or 0.2M trehalose. Freezing of ovarian fragments was slowly performed at 0.3°C/min from the temperature of seeding (-7°C/min) up to -35°C. Thawing, histology, viability tests and electron microscopy evaluation process were performed before and after cryopreservation as described previously.

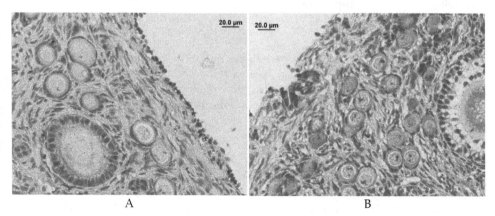

A   B

Fig. 1. Rabbit follicular morphology before (A) and after (B) cryopreservation with PROH and trehalose, with a post-seeding freezing rate at 0.3°C/min

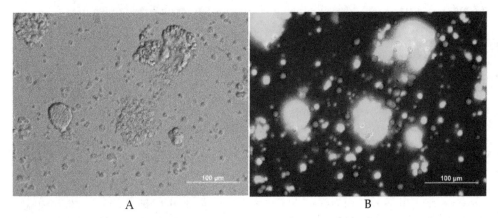

A   B

Fig. 2. View of rabbit isolated follicles under direct light for selection (A) and under fluorescent light (B) after calcein AM/ethidium homodimer I stains to evaluate viability after cryopreservation with PROH, with a post-seeding freezing rate at 0.3°C/min

In control fragments, we observed 72.6 ± 2.8% and 77.7 ± 3.9% of type I follicles (no significant difference) for sucrose and trehalose control groups respectively. After cryopreservation, no statistical difference of the proportions of type I follicles was found between sucrose and trehalose (50.2 ± 4.1% *vs.* 51.1 ± 1.8% respectively) when using DMSO for cryopreservation. When using PROH as permeating CPA, the proportion of type I follicles was lower after cryopreservation with sucrose as compared to trehalose (55.0 ± 3.8% *vs.* 65.0 ± 3.3% respectively; $P<0.05$). When freezing with trehalose the proportion of type I follicles was higher with PROH as compared to DMSO (65.0 ± 3.3% *vs.* 51.1 ± 1.8% respectively; $P<0.01$). Nevertheless, the proportions of type I follicles were significantly reduced after cryopreservation (from $P≤0.01$ to $P<0.001$), whatever the permeating and the non-permeating CPA. No significant difference was observed between the different groups of frozen ovarian cortices, when considering the morphological preservation ratio.

According to these results, the cryopreservation protocol based on a post-seeding freezing rate at 0.3°C/min and using a freezing medium composed of 1.5M PROH, supplemented with 0.2M trehalose was finally evaluated by orthotopic autografting to observe the potential of the cryopreserved follicles to resume follicular growth and to be fertilized.

Before freezing, type II follicles represented the most important part of follicles with morphological defect (19.1 ± 2.9% and 16.1 ± 3.2% in sucrose and trehalose groups respectively). After cryopreservation, follicular defect of type IV (degenerated follicles) was the most important type of morphological defect: 32.5 ± 4.8% and 24.0 ± 1.9% after freezing using DMSO, with sucrose and trehalose respectively; versus 27.2 ± 5.6% and 18.1 ± 3.0% after freezing using PROH, with sucrose and trehalose respectively. The general aspect of ovarian tissue before and after cryopreservation showed a good preservation of structural architecture (follicular structure and connective tissue). Spaces were observed in some case, in the ovarian stroma and the albuginea. Epithelial cells were often absent as compared to the fresh ovarian tissue.

Ultrastructural analysis of the preantral follicles was performed without preliminary selection on semi-thin sections. TEM analysis showed that most follicles of control ovarian tissue had normal ultrastructure, according to mitochondria, nucleus and nuclear membrane, Golgi apparatus and endoplasmic reticulum cisternae observation. They often had vacuoles in cytoplasm and vesicles. Nevertheless, vacuoles were not characteristic of apoptosis. Cellular membranes of the oocyte and follicular cells were in close connection. In general, ovarian stroma was well organised. Fibroblasts and collagen fibres were distinguishable (Fig. 3).

After cryopreservation, oocyte ultrastructure appeared to be similar to the control especially for mitochondria, Golgi apparatus, endoplasmic reticulum, interdigital structure between oocyte and follicular cells (Fig.3). Vesicles and vacuoles were rarely observed. Chromatin of the oocyte was diffused and well preserved. Nevertheless, dark follicular cells or follicular cells without any content were most frequently observed, whereas some follicles showed partial or total disruption of their nuclear membrane whatever the evaluated cryprotective solution. The most important damage observed after cryopreservation was the disorganisation of the ovarian stroma (Fig.3). Fibroblasts showed lack of cytoplasm or important vacuolisation. In general, these damages were less frequently observed after cryopreservation using PROH and trehalose.

2.4.2 Investigations in the queen

The experimental variability showed that neither the structural, nor the estimated models of the experimental design were valid when considering the morphological preservation ratio of the follicles or the viability preservation ratio. So, global discrimination of the chemo-physical parameters was not possible. Nevertheless, the influence of the freezing rate after seeding and after -40°C, and the influence of the addition of sucrose in the freezing medium composed of 1.5M CPA were evaluated and analyzed by classical ANOVA test.

Before freezing, ovarian tissue presented $72.2 \pm 3.6\%$ and $83.8 \pm 2.9\%$ of normal follicles (type I) for group 2°C/min and 0.5°C/min post-seeding freezing rate respectively. When freezing with PROH, and whatever the post-seeding freezing rate, proportions of morphologically normal follicles were not significantly reduced after freezing compared to before freezing ($69.2 \pm 9.1\%$ for 2°C/min group vs. 67.4 ± 2.9°C/min for 0.5°C/min group). After freezing with DMSO, and whatever the post-seeding freezing rate, proportions of type I follicles were significantly reduced ($40.8 \pm 6.6\%$ after freezing at 2°C/min and $51.6 \pm 5.1\%$ after freezing at 0.5°C/min; $P<0.05$). Whatever the post-seeding freezing rate, type III defects were the most frequently observed after freezing. General observation of the ovarian tissue showed a good preservation of the ovarian stroma cells and structure after cryopreservation.

Before freezing, ovarian tissue submitted to a free fall into the freezing chamber during the third phase of the freezing process presented $72.2 \pm 3.6\%$ of type I follicles without any difference with samples directly immersed into the liquid nitrogen ($86.8 \pm 2.5\%$). Proportion of normal follicles decreased significantly after cryopreservation except after freezing with PROH according to a free fall into the freezing chamber ($68.2 \pm 9.1\%$ of type I). After freezing using a direct immersion into liquid nitrogen after $-40°C$, and whatever the CPA, proportions of type I follicles were decreased compared to fresh control.

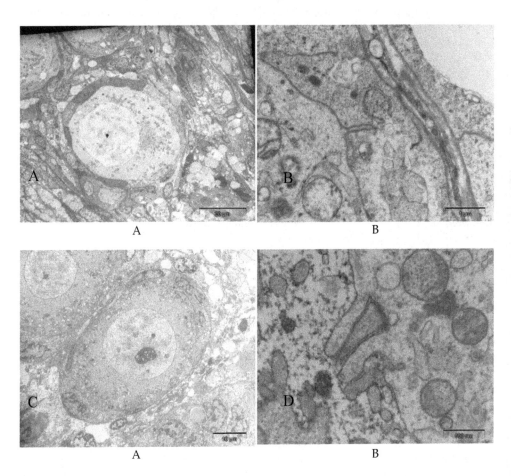

A   B

A   B

Fig. 3. Rabbit follicular ultrastructure before (A & B) and after (C & D) cryopreservation with PROH and trehalose, with a post-seeding freezing rate at $0.3°C/min$

Before freezing, queen ovarian tissue showed $72.2 \pm 3.6\%$ and $74.8 \pm 6.3\%$ of type I follicles respectively for group without and with sucrose without any difference between the two control groups. After freezing, addition of sucrose allowed preserving $63.2 \pm 13.6\%$ of normal follicles versus $68.2 \pm 9.1\%$ without sucrose when associated with PROH, without any significant difference. Contrary to the results after freezing with DMSO, proportion of type I follicles was not significantly different after freezing with PROH compared to fresh control. Morphological defect of type III was the most frequently observed after freezing with PROH. Queen ovarian stroma was well preserved.

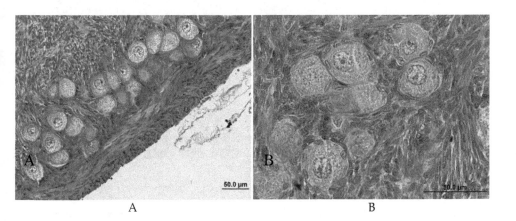

Fig. 4. Queen follicular morphology before (A) and after (B) cryopreservation with 1.5M PROH and 0.2M sucrose. Post-seeding freezing rate at 2°C/min.

In conclusion, queen ovarian tissue seems to be well preserved, without any difference compared to the fresh control when freezing with PROH (according to a free fall in the freezing chamber, without influence of neither sucrose nor post-seeding freezing rate) (Fig. 4).

2.4.3 Investigation in the bitch

In the bitch, the estimated model was validated when the viability preservation ratio was considered. The nature of the non-permeating CPA ($P = 0.37$) did not influence the post thawing viability rate of the ovarian follicles. However all the other factors investigated in this experimental design presented a significant effect on the viability rate. The permeating CPA nature ($P<0.0001$) was the factor that influenced more the viability rate of the follicles. Thus, contrarily to the observations in the other species, the DMSO better preserved the evaluated parameter than PROH. The freezing rate had also a major effect on the viability rate ($P<0.0001$) and a slow freezing rate (0.3°C/min) was less deleterious than the rapid freezing rate for the follicles viability. Moreover the equilibration step also significantly affected the follicles viability, with a beneficial effect of the one step equilibration compared to the 3 steps equilibration. However, no interaction was observed in this model. As a result, the fractional experimental design developed in the bitch, suggested that ovarian tissue should be cryopreserved in a solution containing 2 M DMSO as permeating CPA supplemented with 0.2 M sucrose or trehalose in a one-step equilibration and frozen at a 0.3°C/min freezing rate (Fig. 5).

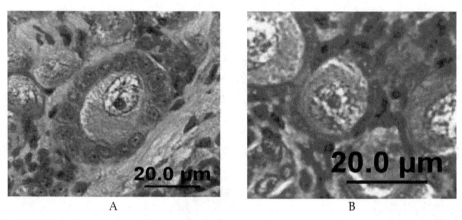

A   B

Fig. 5. Bitch follicular morphology before (A) and after (B) cryopreservation

Consequently, theses parameters validated by the fractional experimental design for viability assessment were used and applied for morphological assessment of frozen-thawed bitch ovarian tissue and comparison with fresh tissue. So, the morphology of preantral follicles was compared between fresh and cryopreserved tissue. The histological analysis revealed that no significant difference was observed between fresh (89.1 ± 6.1 % type I follicles) and cryopreserved ovarian tissue (82.4 ± 4.4 % type I follicles) when type I follicles were observed. The main abnormality observed on preantral follicle after cryopreservation in the bitch was the oocyte nucleus defect (~8%). In this case, the nucleus often appeared pycnotic, with a reduced size and a densely packed chromatin. Sometimes the nuclear membrane was ruptured. However, ooplasm defect were rarely observed alone, but combined with nuclear defects. In some cases few granulosa cells were absent in the primordial or primary follicles. It is probable that ice crystal formation occurring during cooling be responsible of this partially denuded pattern, by destroying or dislodging some granulosa cell during ice expansion.

2.4.4 Investigations in the cow

As for the rabbit doe, the experimental design was valid when considering the morphological preservation ratio of the follicles, but not when considering the viability ratio of the follicles. The concentration of the permeating CPA ($P = 0.59$) and the medium ($P = 0.76$) seem to have no significant effect on the morphological preservation ratio of ovarian follicles. The nature of the permeating CPA seemed to influence the morphological preservation ratio of the follicles ($P = 0.07$) although the non-significant difference. PROH tended to improve the morphological preservation ratio, as compared with DMSO. The nature of the non-permeating CPA ($P = 0.002$) and the cells surfactant ($P = 0.04$) had significant influence. Trehalose and Albumax® improved morphological preservation ratio compared respectively with sucrose and FCS. No significant interaction was observed between the nature of the permeating CPA and its concentration.

In order to discriminate permeating CPAs, ovarian fragments from 5 cows were frozen using 1.5M DMSO or 1.5M PROH with 4 g/L Albumax® and 0.2M trehalose according to a post-seeding freezing rate at 0.3°C/min. Before freezing, ovarian tissue showed 40.6 ± 12.6% of type I preantral follicles. Proportion of type I follicles was reduced to 20.2 ± 3.9% after cryopreservation using DMSO and to 23.8 ± 3.4% when using PROH. No statistical difference was found between DMSO and PROH. Morphological defects of type II were the most important kind of defect (47.0 ± 11.7%). Proportion of type IV follicular defects was significantly improved compared to control for the both CPAs (47.2 ± 7.8% and 44.8 ± 4.4% for DMSO and PROH respectively). Proportion of type III follicles was constant before and after freezing. Ovarian stroma seems to be affected by cryopreservation and shows spaces and disjoined cells.

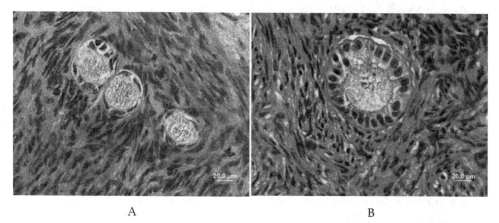

A   B

Fig. 6. Cow follicular morphology before (A) and after (B) cryopreservation using PROH and slow freezing rate

As for the other species, the influence of the post-seeding freezing rate was evaluated when freezing with 2M PROH. Proportions of morphologically normal follicles were significantly reduced after freezing compared with fresh tissue, whatever the post-seeding freezing rate 17.6 ± 6.2% after freezing at 2°C/min vs. 57.8 ± 13.0% before freezing and 17.8 ± 6.5% after freezing at 0.3°C/min vs. 60.0 ± 4.9% before freezing).

2.4.5 In vivo follicle resumption from cryopreserved ovarian tissue

In the doe rabbit, nine pups were born from cryopreserved grafted group, suggesting the efficiency of the cryopreservation protocols based on PROH and trehalose and using very slow freezing rate. At necropsy, follicular structures were observed in most of females.

In the bitch, the cryopreservation method optimized with the fractional experimental design and validated by in vitro assessment (morphology, viability, and toxicity) was then evaluated by heterotopic xenografting to determine whether the ovarian tissue integrity and the follicular growth potential were maintained. The frozen-thawed tissues were grafted to female SCID mice as previously described. The histological assessments of the follicle population revealed a significant increase in the density and distribution of secondary follicles from eight weeks post grafting compared to the follicle population at 1 week (P<0.05). Consequently, the shift from primordial-primary follicles to secondary follicles occurred in a time laps of eight weeks. Moreover morphologically normal follicles were observed until 16 weeks post transplantation and intact secondary follicles with more than 3 layers of granulosa cells and a normal oocyte surrounded by a well-defined zona pellucida were present at this time. Despite a massive follicular loss touching particularly the early follicles and occurring just after grafting, the graft survived long term xenografting. Similarly, after an important loss just after grafting (one week grafts) the stromal cell number increased during the graft period, to reach a density comparable to fresh ovarian tissue at 16 weeks. Otherwise, the vascularization setting-up was assessed by immunohistochemistry as developed previously. The vessel density analysis revealed that already at one week post grafting the vessel density within the graft was comparable to the fresh control ovarian tissue. Moreover, the vessel density tended to increase at 16 weeks post grafting compared to the other groups (fresh control, one and eight weeks, P<0.05) even if no significant difference was registered. No antral follicles was present at the end of the graft time, however, persisting vaginal cornification was noticed on the recipients from 5-9 days post grafting, and estrus behavior was observed several times during the graft period in the recipients cages indicating an hormonal activity resulting from the graft. Taken together, these in vivo results confirm the good preservation of the bitch ovarian tissue by applying our cryopreservation method.

2.5 Discussion

The greater challenge of studies related to the cryopreservation of the ovarian tissue is to define a freezing protocol adapted to the different cell types such as oocytes, follicular cells, stroma cells, etc. One of the objectives of our team was to compare the effects of different freezing parameters based on the morphology and the viability of the follicles, the evaluation of the ultrastructure of the ovarian tissue, the DNA fragmentation of the oocytes or the graft of the ovarian tissue. When the mathematical model was validated, the use of experimental fractional designs allowed us to know simultaneously the individual and the relative effects of different chemo-physical freezing parameters for each species. This statistical method firstly allows a global evaluation of cryopreservation protocols and discriminates the most valuable factors. Finally, the factors which seemed to have a discriminating effect on follicular morphological preservation were evaluated in a wider population. (Neto et al., 2008).

The results of the experimental design in the doe rabbit and in the bitch show that the post-seeding freezing rate is one of the most important chemophysical factors influencing the morphology or the viability of ovarian follicles. A slow freezing rate (0.3°C/min) seems to be more appropriate for the cryopreservation of the doe rabbit and bitch ovarian tissue. Nevertheless, no influence of the freezing rate was observed in the queen and in the cow. Most of the authors use a very slow freezing rate, which is derived from embryo freezing protocols. However, few studies show the importance of this freezing parameter. In contrast to our results in the bitch and in the doe rabbit, but in accordance with our result in the queen and in the cow, Demirci *et al.* observed a high (but not significant) proportion of follicles without any morphological defect after a post-seeding freezing rate of 2°C/min in the ewe (Demirci et al., 2001). Nevertheless, Gook *et al.* also observed a better follicular preservation when using a slow freezing rate (0.3°C/min) with human ovarian tissue (Gook et al., 1999). Whereas Cleary *et al.* observed no difference in terms of follicular growth after grafting, between a conventional embryo freezing protocol (0.3°C/min) and a passive cooling at 1°C/min from 0°C to -84°C on the mouse ovarian tissue (Cleary et al., 2001). Although these two cooling rates (0.3°C/min and 2°C/min) could be considered as slow, these results may be explained by a difference in cell dehydration during the post-seeding step. With rapid cooling rates, we can hypothesise that time required for the exosmose of the cell water is insufficient and consequently promotes the formation of lethal intracellular ice. While at very slow cooling rates, high level of dehydration occurs with concomitant increasing in solute concentration (salting out). Investigations on the freezing rate were extended in the queen with the evaluation of the third freezing phase. When associated with PROH, slow cooling in solid phase seems to be more appropriate. Nevertheless, several authors use cryopreservation protocols with a direct immersion into liquid nitrogen after – 40°C, such as Rodrigues et al. (2004) in the goat, Lucci et al. (2004) in the zebu cow, Santos et al. (2006) in the ewe or Lima et al. (2006) in the cat. Births had been obtained after graft of ovarian fragments frozen with such a protocol in the rabbit doe (Almodin et al., 2004b) and in the ewe (Almodin et al., 2004a), but not in the cat where in vivo follicular growth were observed when using a freezing protocol with controlled third phase (Bosch et al., 2004).

The experimental designs revealed a crucial role of the permeating CPA in the doe rabbit, in the cow and in the bitch, added to a crucial role of the non permeating CPA in the doe rabbit and in the cow. Among the various freezing protocols described in the literature, those using DMSO or PROH as permeating CPA seems to be more efficient, whatever the species. Our results suggest that PROH improve the follicular quality after freezing in the doe rabbit and in the queen. Results of experimental design obtained in the cow suggested a better morphological preservation rate when using PROH. However, in this species, standard comparison between DMSO and PROH doesn't confirm these results. Contrary to that, bitch ovarian tissue seems to be better preserved in freezing medium composed of DMSO. These results were confirmed by the follicular growth observed after xenograft. It can be hypothesized that DMSO penetrate better within the tissue than PROH. Indeed, the bitch ovarian tissue as the goat or the ewe is rich in collagenous fibbers and more fibrous than the doe rabbit ovarian tissue for example. Therefore, a good ability to penetrate within the ovarian tissue is an important characteristic for the chosen CPA Nevertheless, both CPA have sensibly the same molecular weight (PROH: 76.10 g/mol, DMSO: 78.14 g/mol) with a lower weight for PROH. Thus a better penetration of the DMSO cannot be explained by this physical parameter. The explanation may come from the toxicity of both CPAs. Our team also investigated the toxicity of the both CPAs on bitch ovarian tissue after equilibration steps at room temperature without freezing and registered a deleterious effect of PROH compared to DMSO on preantral follicle viability in this species.

Except in canine and feline models, addition of non-permeating CPA in the freezing medium seems to improve the protective effect of CPAs. The protective effect of PROH seems to be improved when it is associated with trehalose, in the doe rabbit and in the cow. This observation was confirmed by electron microscopy evaluation of the doe rabbit ovarian tissue subjected to cryopreservation. Ultrastructure of doe rabbit follicles after cryopreservation was well preserved, but stromal cells and fibroblasts were damaged. Such alterations have been observed in human tissues after cryopreservation (Navarro-Costa et al., 2005; Santos et al., 2010). However, fibroblasts can easily be reproduced by cell division, indicating that damage to the stroma can be repaired. Collagen fibers did not seem to be damaged by cryopreservation in this study, but they were sparse in the doe rabbit ovary. This observation may explain the fragility of doe rabbit ovarian tissue during the equilibration, freezing and thawing steps (Neto et al., 2005).

Sugar are not systematically associated with permeating CPA, but Marsella et al. (2008), showed the advantageous effect of sucrose. Trehalose has been frequently used in embryo cryopreservation, but not in ovarian tissue cryopreservation. Sucrose and trehalose share the property to stabilise cellular membranes and proteins via the formation of hydrogen bonds with polar residues of phospholipidic membrane. This property allows preserving the membrane integrity under anhydrous conditions. Moreover, it modifies the temperature at which the separation of lipid phase occurs during cooling (Crowe et al., 1984; Crowe et al., 1985; Crowe et al., 2001). As compared to other sugars, trehalose seems to have a higher capacity to preserve biomolecules, cellular membrane and cells in a drying or in a freezing state (Crowe et al., 1996; Storey et al., 1998; Sano et al., 1999; Welsh and Herbert, 1999).

Few comparable studies have been reported in the cryopreservation of different mammalian species. Despite encouraging results in the different studied species, and except in the queen, none of evaluated protocols allows preserving the same proportion of normal follicles than before freezing. Most of authors observed a reduction of normal follicles in frozen/thawed ovarian tissue compared with fresh control when using similar freezing protocols in the mouse (Candy et al., 1997), the goat (Rodrigues et al., 2004), the cow (Lucci et al., 2004), and the ewe (Demirci et al., 2002). As for the queen, no morphological difference was observed in human follicles before and after cryopreservation (Hovatta et al., 1996; Fabbri et al., 2006) Newton observed similar proportions of "viable" follicles after freezing when using DMSO, ethylene glycol or PROH and xenografting (Newton et al., 1996).

Live births in the rabbit doe and follicular growths observed in the bitch after grafting of cryopreserved ovarian tissue shows the efficacy of evaluated freezing protocols. Almodin et al. obtained live offspring after grafting of small fragments of cryopreserved rabbit ovarian tissue using 1.5 M DMSO and a very slow post seeding freezing rate (Almodin et al., 2004b). In the bitch, results about in vivo growth obtained after cryopreservation are relatively scarce compared to other species. Ishijima et al. (2006) tried to transplant vitrified ovarian tissue (2M DMSO, 3M PROH, 1M acetamide) into immunodeficient mice during 4 weeks and observed signs of growth in the early follicles (primordial-primary follicles). However, they noticed an important follicular loss occurring just after grafting which is in accordance with our results obtained with slow-frozen tissue. The time necessary for setting up of the neovascularization within the grafted tissue seems to be more deleterious for the cells than the cryopreservation technique itself. Except our results obtained on the bitch ovarian tissue cryopreservation no other results using slow freezing of female germ cells was obtained in this species. Furthermore, live birth has not yet been obtained after ovarian tissue cryopreservation, but the difficulties to mastered *in vitro* maturation and fertilization steps in canines do not contribute to the development of this technique.

For the first time a complete evaluation process of important factors influencing the morphology and the viability of preantral follicles has been performed after equilibration process and freezing in different species. These results suggest that cryopreservation of ovarian tissue is a promising and suitable technique that could be used as complementary tool to embryo cryopreservation, to preserve the animals' genetic resources by the female pathway. Doe rabbit could also be used as a biomedical model to investigate the long term consequences of cryopreservation on ovarian follicles and the birth of future progenies.

3. Perspectives

In definitive, the use of factorial fractional experimental design approach allowed us to develop suitable cryopreservation protocol in different species, while reducing the number of experiments and increasing the number of parameters evaluated. However it can be noticed in our model species, but also in the literature, that results can be radically different according to the species. Moreover, among the numerous articles published on ovarian tissue cryopreservation heterogeneous results can be observed in the same species after applying roughly or widely different slow-freezing protocols. One of the candidates to explain such disparity in the obtained results is the amount of ice crystals formed during slow freezing. Indeed, the strategy of slow cooling is to decrease cell temperature enough slowly to allow removal of most of the freezable intracellular water before reaching the ice nucleation temperature. The main objective of this method is to avoid intracellular ice crystal formation which is known to be lethal. However ice crystallization still occurs extracellularly with the risk of tissue shrinkage or disorganization of the tissue components.

As ice formation and melting are exothermic and endothermic phenomena respectively, they can be objectified and studied by thermodynamical measures. Among the various physical methods of analysis, Differential Scanning Calorimetry (DSC) is an interesting tool. In fact, DSC gives the opportunity to measure important parameters of a cryopreservation solution under dynamic conditions. A cryopreservation solution can thus be characterized by its thermal properties such as temperatures of phase transitions and quantity of ice crystallized and melted. Two types of DSC are commonly used: power-compensation DSC and heat-flux DSC. Our team chose the first one in order to study cryopreservation solutions with a more fundamental approach than with biological methods.

The power compensation DSC is based on the "zero balance principle" as explained as follow. The sample and a reference are placed in two microfurnaces which are continuously cooled by liquid nitrogen. The temperature of each microfurnace can be, on the one hand, precisely measured by a temperature sensor and, on the other hand, precisely adjusted by a heating resistor. Each microfurnace contains one sensor and one resistor. The principle of the power compensation DSC is to maintain the two microfurnaces under the same temperature regardless of phase transitions or reactions occurring in the sample. Thus, when a phase transition occurs in the sample, the heat released or absorbed by the sample has to be compensated by the heating resistor which is below the sample. Consequently, the calorimeter measures a difference between the heating powers provided by the two resistors. This difference reveals the phase transition. When this phase transition is crystallization, this difference allows also us to measure the quantity of ice formed.

Since recent years, several strategies are developed to avoid deleterious effects of the ice crystal formation during cooling, and thermodynamical approaches are more and more associated to these strategies.

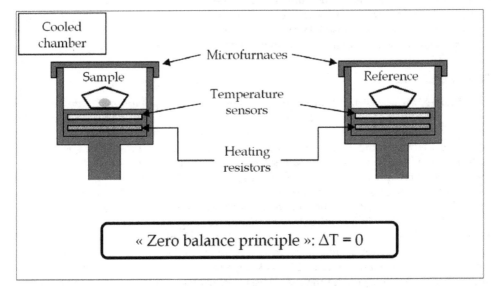

Fig. 7. Scheme of power-compensation DSC

The specimen and reference temperatures are controlled independently using separate (identical) ovens. The temperature difference between the sample and reference is maintained to zero by varying the power input to the two furnaces. This energy is then a measure of the enthalpy or hat capacity changes in the test specimen (relative to the reference).

3.1 Prevent the ice formation

The physical definition of vitrification is the glass-like solidification of solutions at low temperatures, without the formation of intracellular ice crystals. The vitrification technique is the solidification of a liquid without ice crystal formation. This phase is obtained by increasing the solute concentration of the vitrification media (increased viscosity that makes water solidify without expansion) or by using very fast cooling rate to avoid molecular rearrangement into ice crystals but into an amorphous glass. (Vajta, 2000). Consequently, the risk of injuries due to intra- or extracellular ice crystallization is avoided, which constitutes the main advantage of this technique.

Despite the fact that slow freezing is the most widely used cryopreservation technique, vitrification is a viable and promising alternative that is increasingly becoming more attractive to the commercial sector. Vitrification has been used for oocyte and tissue cryopreservation since the 1980s (Rall and Fahy, 1985). Since the first use of this approach in ovarian tissue cryopreservation, it has been well optimized, particularly by decreasing the concentration of CPAs used to reduce their toxicity, but also by increasing the cooling rate applied (Chen et al., 2006; Keros et al., 2009). These improvements of vitrification in ovarian tissue but also in embryo and oocyte vitrification place the vitrification as a gold standard method for germ cells cryopreservation (Saragusty and Arav, 2011). It has been suggested that with time, conventional slow freezing will be replaced entirely by vitrification techniques (Vajta and Kuwayama, 2006).

As mentioned above, DSC can be a precious aid for the study of vitrification. In fact, DSC allows the detection of phase transitions, both crystallization or glass transition, and can thus determine three thermal properties of vitrification solutions: the vitreous transition temperature, the critical cooling rate and the critical warming rate (Baudot et al., 2007). These three thermal properties are specific to each vitrification solution and could be used for a better utilization of the solutions. The critical cooling rate and the critical warming rate are the cooling and warming speeds above which the crystallization cannot occur between the vitreous transition temperature and the crystallization temperature. The calculation of these critical cooling and warming rates is based on a semi-empiric model developed by Boutron in 1986 according to the "classic" theory of crystallization (Boutron, 1986). This model reproduces well analytically the experimental results, but some approximations are introduced.

3.2 Limit the ice formation

As described in the introductive part, extracellular crystallization occurs during slow freezing. The penetrating CPAs like DMSO, PROH or ethylene glycol, are mainly used to reduce the volume of freezable ice within the cells, but their respective toxicity for the cells limit the potential concentration usable. However, since the last decades, non-penetrating CPAs like disaccharides have been widely used in cryoprotective solution to improve cell preservation and reduce the penetrating CPA amount necessary. To our knowledge, the first use of sucrose as cryoprotectant of ovarian tissue was in 1996 with human ovaries (Hovatta et al., 1996). Sugars have been proved to play a role on ice crystal formation and stability (Kuleshova et al., 1999). Previous studies on mouse embryos revealed that sugars increase the homogeny ice transition temperature in freezing solutions. By this way, dehydration of the cells occurs when they are still permeable to water without reaching ice nucleation temperature. Moreover, the trehalose was reported to reduce the size of ice crystals and shorten their elongation during freezing (Sei et al., 2002).

On the basis of the contribution of DSC in other fields of cryopreservation [vitrification of ovary (Baudot et al., 2007), slow-freezing or vitrification of plants (Volk and Walters, 2006; Skyba et al., 2011), slow-freezing of aquatic crustaceans (Issartel et al., 2006)], our team chose a thermodynamic approach to study slow-freezing solutions for ovarian tissue. In fact, DSC allows the measure of the maximal quantity of ice formed in a solution (Q_{max}). Q_{max} corresponds to the definition of Boutron : "The heat of solidification are represented by the numbers q of grams of ice whose solidification at 0°C would liberate the same amount of heat as that from 100g of solution on crossing the corresponding peaks. They are close to the real quantities of ice crystallized in % (w/w) of the solution when it is ice which crystallizes. One obtains the heat in calories per 100g of solution by multiplying q by 79.78" (Boutron, 1984). Our team has thus quantified the quantity of ice formed in solutions used for the cryopreservation of ovarian tissue of different species (table 6).

Species	Q_{max} in percentage (w/w) of solution
Bitch	36.81
Doe rabbit	45.79
Cow	37.64

Table 6. Maximal quantity of ice crystallized (Q_{max}) in solutions used for the cryopreservation of ovarian tissue of bitch, doe rabbit and cow.

Then, our team decided to explore two research areas. On the one hand, DSC can compare the quantity of ice formed in two different cryopreservation solutions. Thus, it seems possible to select the most suitable cryopreservation solutions for slow freezing methods. The first results obtained in doe rabbit seem to confirm this hypothesis. In fact between two cryopreservation solutions tested, those for which the quantity of ice formed was the lowest, was also the one with the best biological results. On the other hand, for a given solution, DSC allows the measure of the quantity of ice formed for different freezing kinetics. Consequently, it seems also possible to select the most suitable kinetics according to the cryopreservation solutions.

3.3 Promote the formation of a non-vulnerable extracellular ice

Another approach to optimize cryopreservation process should be to control the ice crystal growth and shape in order to promote the less deleterious crystallization. It is already assumed that intracellular ice formation is lethal. Nonetheless, the recent observations of rabbit ovarian tissue by cryoscanning electronic microscopy and cryofracture reveal that depending on the cooling rate, the extracellular ice shape is modified. Moreover, according to these results the seeding temperature influences the shape and regularity of the ice crystals resulting in large uniform crystals when seeding was induced close to the solution melting point (Gosden et al., 2010).

Otherwise, a better understanding of intracellular ice formation can also be advantageous to improve cryopreservation processes. Indeed, Han *et al* (2009) investigated the size of intracellular ice crystals in mouse oocytes by cryomicroscopy. They conclude that increasing the concentration of macromolecules in the cells by increasing the extracellular non permeating solute concentration significantly reduced the required permeating CPA concentration for intracellular vitrification. Moreover they also observed that intracellular ice melting point was always lower than extracellular ice. Taken together, this information can be helpful to optimize the cryopreservation protocols.

Regarding DSC, a recent study showed that it is possible to differentiate the crystallization of intra and extracellular ice depending on freezing kinetics (Seki et al., 2009). Consequently, in addition to the measure of the quantity of crystallized ice, DSC can provide a better control of ice formation.

4. Acknowledgment

The results presented in this chapter were partially supported financially by Région Rhône-Alpes and the breeding center (CESECAH) of the French Guide Dog Federation (FFAC). The authors also thank the LLC Hycole (Marcoing, France) for their technical support.

The authors want to thank equally regards the Electronic Microscopy Center of the University of Lyon, the Laboratory of Pathological Anatomy of the Veterinary Campus of

Lyon and the Institute Claude Bourgelat for access to their facilities. The authors are also grateful to all veterinarians which allowed the ovary collection.

5. References

Abrishami, M.; Anzar, M.; Yang, Y.&Honaramooz, A. (2010). Cryopreservation of immature porcine testis tissue to maintain its developmental potential after xenografting into recipient mice. *Theriogenology*, Vol.73, No.1, pp: 86-96,

Almodin, C. G.; Minguetti-Camara, V. C.; Meister, H.; Ceschin, A. P.; Kriger, E.&Ferreira, J. O. (2004a). Recovery of natural fertility after grafting of cryopreserved germinative tissue in ewes subjected to radiotherapy. *Fertil Steril*, Vol.81, No.1, pp: 160-164,

Almodin, C. G.; Minguetti-Camara, V. C.; Meister, H.; Ferreira, J. O.; Franco, R. L.; Cavalcante, A. A.; Radaelli, M. R.; Bahls, A. S.; Moron, A. F.&Murta, C. G. (2004b). Recovery of fertility after grafting of cryopreserved germinative tissue in female rabbits following radiotherapy. *Hum Reprod*, Vol.19, No.6, pp: 1287-1293,

Andrabi, S. M.&Maxwell, W. M. (2007). A review on reproductive biotechnologies for conservation of endangered mammalian species. *Anim Reprod Sci*, Vol.99, No.3-4, pp: 223-243,

Aubard, Y.; Newton, H.; Scheffer, G.&Gosden, R. (1998). Conservation of the follicular population in irradiated rats by the cryopreservation and orthotopic autografting of ovarian tissue. *Eur J Obstet Gynecol Reprod Biol*, Vol.79, No.1, pp: 83-87,

Bahadur, G.&Ralph, D. (1999). Gonadal tissue cryopreservation in boys with paediatric cancers. *Hum Reprod*, Vol.14, No.1, pp: 11-17,

Bahadur, G.; Chatterjee, R.&Ralph, D. (2000). Testicular tissue cryopreservation in boys. Ethical and legal issues: case report. *Hum Reprod*, Vol.15, No.6, pp: 1416-1420,

Baudot, A.; Courbiere, B.; Odagescu, V.; Salle, B.; Mazoyer, C.; Massardier, J.&Lornage, J. (2007). Towards whole sheep ovary cryopreservation. *Cryobiology*, Vol.55, No.3, pp: 236-248,

Boettcher, P. J.; Stella, A.; Pizzi, F.&Gandini, G. (2005). The combined use of embryos and semen for cryogenic conservation of mammalian livestock genetic resources. *Genet Sel Evol*, Vol.37, No.6, pp: 657-675,

Bosch, P.; Hernandez-Fonseca, H. J.; Miller, D. M.; Wininger, J. D.; Massey, J. B.; Lamb, S. V.&Brackett, B. G. (2004). Development of antral follicles in cryopreserved cat ovarian tissue transplanted to immunodeficient mice. *Theriogenology*, Vol.61, No.2-3, pp: 581-594,

Boutron, P. (1984). More accurate determination of the quantity of ice crystallised at low cooling rates in glycerol and 1,2-propanediol aqueous solutions : comparison with equilibrium. *Cryobiology*, Vol.21, 183-191,

Boutron, P. (1986). Comparison with the theory of the kinetics and extent of ice crystallization and of glass-forming tendency in aqueous cryoprotective solutions. *Cryobiology*, Vol.23, 88-102,

Candy, C. J.; Wood, M. J.&Whittingham, D. G. (1997). Effect of cryoprotectants on the survival of follicles in frozen mouse ovaries. *J Reprod Fertil*, Vol.110, No.1, pp: 11-19,

Checura, C. M.&Seidel, G. E., Jr. (2007). Effect of macromolecules in solutions for vitrification of mature bovine oocytes. *Theriogenology*, Vol.67, No.5, pp: 919-930,

Chen, S. U.; Chien, C. L.; Wu, M. Y.; Chen, T. H.; Lai, S. M.; Lin, C. W.&Yang, Y. S. (2006). Novel direct cover vitrification for cryopreservation of ovarian tissues increases follicle viability and pregnancy capability in mice. *Hum Reprod*, Vol.21, No.11, pp: 2794-2800,

Cleary, M.; Snow, M.; Paris, M.; Shaw, J.; Cox, S. L.&Jenkin, G. (2001). Cryopreservation of mouse ovarian tissue following prolonged exposure to an Ischemic environment. *Cryobiology*, Vol.42, No.2, pp: 121-133,

Commin, L.; Rosset, E.; Allard, A.; Joly, T.; Guérin, P.; Neto, V.&Buff, S. (2011). Assesment of tissue functionality after xenotransplantation of canine ovarian cortex to severe combined immunodeficient mice (SCID mice), *Theriogenology Conference and Symposia*, pp: Milwaukee, Wisconsin

Critser, J. K. (1998). Current status of semen banking in the USA. *Hum Reprod*, Vol.13 Suppl 2, 55-67; discussion 68-59,

Crowe, J. H.; Crowe, L. M.&Chapman, D. (1984). Preservation of membranes in anhydrobiotic organisms: the role of trehalose. *Science*, Vol.223, No.4637, pp: 701-703,

Crowe, J. H.; Crowe, L. M.; Oliver, A. E.; Tsvetkova, N.; Wolkers, W.&Tablin, F. (2001). The trehalose myth revisited: introduction to a symposium on stabilization of cells in the dry state. *Cryobiology*, Vol.43, No.2, pp: 89-105,

Crowe, L. M.; Reid, D. S.&Crowe, J. H. (1996). Is trehalose special for preserving dry biomaterials? *Biophys J*, Vol.71, No.4, pp: 2087-2093,

Crowe, L. M.; Crowe, J. H.; Rudolph, A.; Womersley, C.&Appel, L. (1985). Preservation of freeze-dried liposomes by trehalose. *Arch Biochem Biophys*, Vol.242, No.1, pp: 240-247,

Curaba, M.; Verleysen, M.; Amorim, C. A.; Dolmans, M. M.; Van Langendonckt, A.; Hovatta, O.; Wyns, C.&Donnez, J. (2011). Cryopreservation of prepubertal mouse testicular tissue by vitrification. *Fertil Steril*, Vol.95, No.4, pp: 1229-1234 e1221,

Deanesly, R. (1954). Immature rat ovaries grafted after freezing and thawing. *J Endocrinol*, Vol.11, No.2, pp: 197-200,

Demirci, B.; Lornage, J.; Salle, B.; Frappart, L.; Franck, M.&Guerin, J. F. (2001). Follicular viability and morphology of sheep ovaries after exposure to cryoprotectant and cryopreservation with different freezing protocols. *Fertil Steril*, Vol.75, No.4, pp: 754-762,

Demirci, B.; Salle, B.; Frappart, L.; Franck, M.; Guerin, J. F.&Lornage, J. (2002). Morphological alterations and DNA fragmentation in oocytes from primordial and primary follicles after freezing-thawing of ovarian cortex in sheep. *Fertil Steril*, Vol.77, No.3, pp: 595-600,

Donnez, J.; Dolmans, M. M.; Demylle, D.; Jadoul, P.; Pirard, C.; Squifflet, J.; Martinez-Madrid, B.&van Langendonckt, A. (2004). Livebirth after orthotopic transplantation of cryopreserved ovarian tissue. *Lancet*, Vol.364, No.9443, pp: 1405-1410,

Edge, B.; Holmes, D.&Makin, G. (2006). Sperm banking in adolescent cancer patients. *Arch Dis Child*, Vol.91, No.2, pp: 149-152,

Fabbri, R.; Pasquinelli, G.; Bracone, G.; Orrico, C.; Di Tommaso, B.&Venturoli, S. (2006). Cryopreservation of human ovarian tissue. *Cell Tissue Bank*, Vol.7, No.2, pp: 123-133,

FAO (2007). The state of the world's animal genetic resources for food and agriculture, *Food and Agriculture Organization of the United Nations*, Commission on Genetic Resources for Food and Agriculture, Rome, Italy

Fishel, S.; Green, S.; Hunter, A.; Lisi, F.; Rinaldi, L.; Lisi, R.&McDermott, H. (1997). Human fertilization with round and elongated spermatids. *Hum Reprod*, Vol.12, No.2, pp: 336-340,

Gook, D. A.; Edgar, D. H.&Stern, C. (1999). Effect of cooling rate and dehydration regimen on the histological appearance of human ovarian cortex following cryopreservation in 1, 2-propanediol. *Hum Reprod*, Vol.14, No.8, pp: 2061-2068,

Gook, D. A.; Edgar, D. H.&Stern, C. (2004). Cryopreservation of human ovarian tissue. *Eur J Obstet Gynecol Reprod Biol*, Vol.113 Suppl 1, S41-44,

Gosden, R. G.; Baird, D. T.; Wade, J. C.&Webb, R. (1994). Restoration of fertility to oophorectomized sheep by ovarian autografts stored at -196 degrees C. *Hum Reprod*, Vol.9, No.4, pp: 597-603,

Gosden, R. G.; Yin, H.; Bodine, R. J.&Morris, G. J. (2010). Character, distribution and biological implications of ice crystallization in cryopreserved rabbit ovarian tissue revealed by cryo-scanning electron microscopy. *Hum Reprod*, Vol.25, No.2, pp: 470-478,

Gougeon, A.&Chainy, G.B.N. (1987). Morphometric studies of small follicles in ovaries of women at different ages. *J Reprod Fertil*, Vol.81, 433-442,

Guerin, J. F. (2005). [Testicular tissue cryoconservation for prepubertal boy: indications and feasibility]. *Gynecol Obstet Fertil*, Vol.33, No.10, pp: 804-808,

Han, X.&Critser, J. K. (2009). Measurement of the size of intracellular ice crystals in mouse oocytes using a melting point depression method and the influence of intracellular solute concentrations. *Cryobiology*, Vol.59, No.3, pp: 302-307,

Holt, W. V.&Pickard, A. R. (1999). Role of reproductive technologies and genetic resource banks in animal conservation. *Rev Reprod*, Vol.4, No.3, pp: 143-150,

Honaramooz, A.; Snedaker, A.; Boiani, M.; Scholer, H.; Dobrinski, I.&Schlatt, S. (2002). Sperm from neonatal mammalian testes grafted in mice. *Nature*, Vol.418, No.6899, pp: 778-781,

Hovatta, O.; Silye, R.; Krausz, T.; Abir, R.; Margara, R.; Trew, G.; Lass, A.&Winston, R. M. (1996). Cryopreservation of human ovarian tissue using dimethylsulphoxide and propanediol-sucrose as cryoprotectants. *Hum Reprod*, Vol.11, No.6, pp: 1268-1272,

Isachenko, V.; Isachenko, E.; Weiss, J. M.; Todorov, P.&Kreienberg, R. (2009). Cryobanking of human ovarian tissue for anti-cancer treatment: comparison of vitrification and conventional freezing. *Cryo Letters*, Vol.30, No.6, pp: 449-454,

Ishijima, T.; Kobayashi, Y.; Lee, D. S.; Ueta, Y. Y.; Matsui, M.; Lee, J. Y.; Suwa, Y.; Miyahara, K.&Suzuki, H. (2006). Cryopreservation of canine ovaries by vitrification. *J Reprod Dev*, Vol.52, No.2, pp: 293-299,

Issartel, J.; Voituron, Y.; Odagescu, V.; Baudot, A.; Guillot, G.; Ruaud, J. P.; Renault, D.; Vernon, P.&Hervant, F. (2006). Freezing or supercooling: how does an aquatic subterranean crustacean survive exposures at subzero temperatures? *J Exp Biol*, Vol.209, No.Pt 17, pp: 3469-3475,

Jahnukainen, K.; Ehmcke, J.; Hergenrother, S. D.&Schlatt, S. (2007). Effect of cold storage and cryopreservation of immature non-human primate testicular tissue on spermatogonial stem cell potential in xenografts. *Hum Reprod*, Vol.22, No.4, pp: 1060-1067,

Jezek, D.; Schulze, W.; Kalanj-Bognar, S.; Vukelic, Z.; Milavec-Puretic, V.&Krhen, I. (2001). Effects of various cryopreservation media and freezing-thawing on the morphology of rat testicular biopsies. *Andrologia*, Vol.33, No.6, pp: 368-378,

Johnston, L. A.&Lacy, R. C. (1995). Genome resource banking for species conservation: selection of sperm donors. *Cryobiology*, Vol.32, No.1, pp: 68-77,

Karlsson, J.O.M.&Toner, M. (1996). Long-term storage of tissues by cryopreservation : critical issues. *Biomaterials*, Vol.17, 243-256,

Keros, V.; Rosenlund, B.; Hultenby, K.; Aghajanova, L.; Levkov, L.&Hovatta, O. (2005). Optimizing cryopreservation of human testicular tissue: comparison of protocols with glycerol, propanediol and dimethylsulphoxide as cryoprotectants. *Hum Reprod*, Vol.20, No.6, pp: 1676-1687,

Keros, V.; Hultenby, K.; Borgstrom, B.; Fridstrom, M.; Jahnukainen, K.&Hovatta, O. (2007). Methods of cryopreservation of testicular tissue with viable spermatogonia in pre-pubertal boys undergoing gonadotoxic cancer treatment. *Hum Reprod*, Vol.22, No.5, pp: 1384-1395,

Keros, V.; Xella, S.; Hultenby, K.; Pettersson, K.; Sheikhi, M.; Volpe, A.; Hreinsson, J.&Hovatta, O. (2009). Vitrification versus controlled-rate freezing in cryopreservation of human ovarian tissue. *Hum Reprod*, Vol.24, No.7, pp: 1670-1683,

Kuleshova, L. L.; MacFarlane, D. R.; Trounson, A. O.&Shaw, J. M. (1999). Sugars exert a major influence on the vitrification properties of ethylene glycol-based solutions and have low toxicity to embryos and oocytes. *Cryobiology*, Vol.38, No.2, pp: 119-130,

Ledda, S.; Leoni, G.; Bogliolo, L.&Naitana, S. (2001). Oocyte cryopreservation and ovarian tissue banking. *Theriogenology*, Vol.55, No.6, pp: 1359-1371,

Lima, A. K.; Silva, A. R.; Santos, R. R.; Sales, D. M.; Evangelista, A. F.; Figueiredo, J. R.&Silva, L. D. (2006). Cryopreservation of preantral ovarian follicles in situ from domestic cats (Felis catus) using different cryoprotective agents. *Theriogenology*, Vol.66, No.6-7, pp: 1664-1666,

Lucci, C. M.; Kacinskis, M. A.; Lopes, L. H.; Rumpf, R.&Bao, S. N. (2004). Effect of different cryoprotectants on the structural preservation of follicles in frozen zebu bovine (Bos indicus) ovarian tissue. *Theriogenology*, Vol.61, No.6, pp: 1101-1114,

Luvoni, G. C.; Chigioni, S.&Beccaglia, M. (2006). Embryo production in dogs: from in vitro fertilization to cloning. *Reprod Domest Anim*, Vol.41, No.4, pp: 286-290,

Maclellan, L. J.; Carnevale, E. M.; Coutinho da Silva, M. A.; Scoggin, C. F.; Bruemmer, J. E.&Squires, E. L. (2002). Pregnancies from vitrified equine oocytes collected from super-stimulated and non-stimulated mares. *Theriogenology*, Vol.58, No.5, pp: 911-919,

Marsella, T.; Sena, P.; Xella, S.; La Marca, A.; Giulini, S.; De Pol, A.; Volpe, A.&Marzona, L. (2008). Human ovarian tissue cryopreservation: effect of sucrose concentration on morphological features after thawing. *Reprod Biomed Online*, Vol.16, No.2, pp: 257-267,

Massip, A. (2003). Cryopreservation of bovine oocytes: current status and recent developments. *Reprod Nutr Dev*, Vol.43, No.4, pp: 325-330,

Mazur, P.&Schneider, U. (1986). Osmotic responses of preimplantation mouse and bovine embryos and their cryobiological implications. *Cell Biophys*, Vol.8, No.4, pp: 259-285,

Mechakra, A.; Auberger, B.; Remeuf, F.&Lenoir, J. (1999). Optimisation d'un milieu de culture pour la production d'enzymes protéolytiques acides pas *Penicilium camemberti*. *Sciences des aliments*, Vol.19, 663-675,

Milazzo, J. P.; Vaudreuil, L.; Cauliez, B.; Gruel, E.; Masse, L.; Mousset-Simeon, N.; Mace, B.&Rives, N. (2008). Comparison of conditions for cryopreservation of testicular tissue from immature mice. *Hum Reprod*, Vol.23, No.1, pp: 17-28,

Milazzo, J. P.; Travers, A.; Bironneau, A.; Safsaf, A.; Gruel, E.; Arnoult, C.; Mace, B.; Boyer, O.&Rives, N. (2010). Rapid screening of cryopreservation protocols for murine prepubertal testicular tissue by histology and PCNA immunostaining. *J Androl*, Vol.31, No.6, pp: 617-630,

Navarro-Costa, P.; Correia, S. C.; Gouveia-Oliveira, A.; Negreiro, F.; Jorge, S.; Cidadao, A. J.; Carvalho, M. J.&Plancha, C. E. (2005). Effects of mouse ovarian tissue cryopreservation on granulosa cell-oocyte interaction. *Hum Reprod*, Vol.20, No.6, pp: 1607-1614,

Neto, V.; Joly, T.&Salvetti, P. (2007a). Ovarian tissue cryopreservation in the doe rabbit: from freezing to birth *Cryobiology*, Vol.55, No.3, pp: 344,

Neto, V.; Guerin, P.; Lornage, J.; Corrao, N.; Buff, S.&Joly, T. (2005). [Follicular morphology after ovarian cortex cryopreservation in the rabbit doe(Oryctolagus cuniculus)]. *Gynecol Obstet Fertil*, Vol.33, No.10, pp: 793-798,

Neto, V.; Joly, T.; Sailley, F.; Corrao, N.; Guérin, P.&Buff, S. (2006). Assessment of different chemo-physical factors influencing the cryopreservation of queen's ovarian cortex, *5th Biannual EVSSAR Congress, 7-9 April*, pp: 283-284, Budapest, Hungary

Neto, V.; Buff, S.; Lornage, J.; Bottollier, B.; Guerin, P.&Joly, T. (2008). Effects of different freezing parameters on the morphology and viability of preantral follicles after cryopreservation of doe rabbit ovarian tissue. *Fertil Steril*, Vol.29, No.5, pp: 1348-1356,

Neto, V.; Joly, T.; Salvetti, P.; Lefranc, A.C.; Corrao, N.; Guerin, P.&Buff, S. (2007b). Ovarian tissue cryopreservation in the Doe rabbit: from freezing to birth, *44th annual Congress of the Society of Cryobiology*, pp: Lake Louise, Canada

Newton, H.; Aubard, Y.; Rutherford, A.; Sharma, V.&Gosden, R. (1996). Low temperature storage and grafting of human ovarian tissue. *Hum Reprod*, Vol.11, No.7, pp: 1487-1491,

Otoi, T.; Yamamoto, K.; Koyama, N.; Tachikawa, S.&Suzuki, T. (1996). A frozen-thawed in vitro-matured bovine oocyte derived calf with normal growth and fertility. *J Vet Med Sci*, Vol.58, No.8, pp: 811-813,

Parkes, A. S.&Smith, A. U. (1953). Regeneration of rat ovarian tissue grafted after exposure to low temperatures. *Proc R Soc Lond B Biol Sci*, Vol.140, No.901, pp: 455-470,

Parkes, A. S.&Smith, A. U. (1954). Preservation of ovarian tissue at -79 degrees C. for transplantation. *Acta Endocrinol (Copenh)*, Vol.17, No.1-4, pp: 313-320,

Parrott, D. M. (1960). The fertility of mice with orthotopic ovarian grafts from frozen tissue. *J. Reprod. Fertil*, Vol.1, 230-241,

Paynter, S. J.; Cooper, A.; Fuller, B. J.&Shaw, R. W. (1999). Cryopreservation of bovine ovarian tissue: structural normality of follicles after thawing and culture in vitro. *Cryobiology*, Vol.38, No.4, pp: 301-309,

Pereira, R. M.&Marques, C. C. (2008). Animal oocyte and embryo cryopreservation. *Cell Tissue Bank*, Vol.9, No.4, pp: 267-277,

Pullium, J. K.; Milner, R.&Tuma, G. A. (2008). Pregnancy following homologous prepubertal ovarian transplantation in the dog. *J Exp Clin Assist Reprod*, Vol.5, No.1, pp: 1,

Rall, W. F.&Fahy, G. M. (1985). Ice-free cryopreservation of mouse embryos at -196 degrees C by vitrification. *Nature*, Vol.313, No.6003, pp: 573-575,

Revel, A.&Mejia, J. (2010). Testicular tissue cryopreservation, In: Testicular tissue cryopreservation, In: Fertility cryopreservation, (Ed.)^(Eds), Cambridge University Press, 978-0-521-51778-2, United Kingdom

Rodrigues, A. P.; Amorim, C. A.; Costa, S. H.; Matos, M. H.; Santos, R. R.; Lucci, C. M.; Bao, S. N.; Ohashi, O. M.&Figueiredo, J. R. (2004). Cryopreservation of caprine ovarian tissue using dimethylsulphoxide and propanediol. *Anim Reprod Sci*, Vol.84, No.1-2, pp: 211-227,

Salvetti, P.; Buff, S.; Afanassieff, M.; Daniel, N.; Guerin, P.&Joly, T. (2010). Structural, metabolic and developmental evaluation of ovulated rabbit oocytes before and after cryopreservation by vitrification and slow freezing. *Theriogenology*, Vol.74, No.5, pp: 847-855,

Sano, F.; Asakawa, N.; Inoue, Y.&Sakurai, M. (1999). A dual role for intracellular trehalose in the resistance of yeast cells to water stress. *Cryobiology*, Vol.39, No.1, pp: 80-87,

Santos, R. R.; Rodrigues, A. P.; Costa, S. H.; Silva, J. R.; Matos, M. H.; Lucci, C. M.; Bao, S. N.; van den Hurk, R.&Figueiredo, J. R. (2006). Histological and ultrastructural analysis of cryopreserved sheep preantral follicles. *Anim Reprod Sci*, Vol.91, No.3-4, pp: 249-263,

Santos, R. R.; Amorim, C.; Cecconi, S.; Fassbender, M.; Imhof, M.; Lornage, J.; Paris, M.; Schoenfeldt, V.&Martinez-Madrid, B. (2010). Cryopreservation of ovarian tissue: an emerging technology for female germline preservation of endangered species and breeds. *Anim Reprod Sci*, Vol.122, No.3-4, pp: 151-163,

Saragusty, J.&Arav, A. (2011). Current progress in oocyte and embryo cryopreservation by slow freezing and vitrification. *Reproduction*, Vol.141, No.1, pp: 1-19,

Schlatt, S.; Kim, S. S.&Gosden, R. (2002). Spermatogenesis and steroidogenesis in mouse, hamster and monkey testicular tissue after cryopreservation and heterotopic grafting to castrated hosts. *Reproduction*, Vol.124, No.3, pp: 339-346,

Sei, T.; Gonda, T.&Arima, Y. (2002). Growth rate and morphology of ice crystals growing in a solution of trehalose and water. *Journal of Crystal Growth*, Vol.240, 218-229,

Seki, S.; Kleinhans, F. W.&Mazur, P. (2009). Intracellular ice formation in yeast cells vs. cooling rate: predictions from modeling vs. experimental observations by differential scanning calorimetry. *Cryobiology*, Vol.58, No.2, pp: 157-165,

Sherman, J. K.&Bunge, R. G. (1953). Effect of glycerol and freezing on some staining reactions of human spermatozoa. *Proc Soc Exp Biol Med*, Vol.84, No.1, pp: 179-180,

Skyba, M.; Faltus, M.; Zàmecnik, J.&Cellàrovà, E. (2011). Thermal analysis of cryopreserved Hypericum perforatum L. shoot tips: Cooling regime dependent dehydration and ice growth. *Thermochimica acta*, Vol.514, 22-27,

Smith, A.U. (1952). Culture of ovarian granulosa cells after cooling to very low temperatures. *Exp. Cell. Res.*, Vol.3, 574-583,

Sofikitis, N.; Mantzavinos, T.; Loutradis, D.; Yamamoto, Y.; Tarlatzis, V.&Miyagawa, I. (1998). Ooplasmic injections of secondary spermatocytes for non-obstructive azoospermia. *Lancet*, Vol.351, No.9110, pp: 1177-1178,

Stahler, E.; Sturm, G.; Spatling, L.; Daume, E.&Buchholz, R. (1976). Investigations into the preservation of human ovaries by means of a cryoprotectivum. *Arch Gynakol*, Vol.221, No.4, pp: 339-344,

Storey, B. T.; Noiles, E. E.&Thompson, K. A. (1998). Comparison of glycerol, other polyols, trehalose, and raffinose to provide a defined cryoprotectant medium for mouse sperm cryopreservation. *Cryobiology*, Vol.37, No.1, pp: 46-58,

Travers, A.; Milazzo, J. P.; Perdrix, A.; Metton, C.; Bironneau, A.; Mace, B.&Rives, N. (2011). Assessment of freezing procedures for rat immature testicular tissue. *Theriogenology*, Vol.76, No.6, pp: 981-990,

Vajta, G. (2000). Vitrification of the oocytes and embryos of domestic animals. *Anim Reprod Sci*, Vol.60-61, 357-364,

Vajta, G.&Kuwayama, M. (2006). Improving cryopreservation systems. *Theriogenology*, Vol.65, No.1, pp: 236-244,

Vanderzwalmen, P.; Zech, H.; Birkenfeld, A.; Yemini, M.; Bertin, G.; Lejeune, B.; Nijs, M.; Segal, L.; Stecher, A.; Vandamme, B.; van Roosendaal, E.&Schoysman, R. (1997). Intracytoplasmic injection of spermatids retrieved from testicular tissue: influence of testicular pathology, type of selected spermatids and oocyte activation. *Hum Reprod*, Vol.12, No.6, pp: 1203-1213,

Volk, G. M.&Walters, C. (2006). Plant vitrification solution 2 lowers water content and alters freezing behavior in shoot tips during cryoprotection. *Cryobiology*, Vol.52, No.1, pp: 48-61,

Welsh, D. T.&Herbert, R. A. (1999). Osmotically induced intracellular trehalose, but not glycine betaine accumulation promotes desiccation tolerance in Escherichia coli. *FEMS Microbiol Lett*, Vol.174, No.1, pp: 57-63,

Woods, E. J.; Benson, J. D.; Agca, Y.&Critser, J. K. (2004). Fundamental cryobiology of reproductive cells and tissues. *Cryobiology*, Vol.48, No.2, pp: 146-156,

Wyns, C.; Van Langendonckt, A.; Wese, F. X.; Donnez, J.&Curaba, M. (2008). Long-term spermatogonial survival in cryopreserved and xenografted immature human testicular tissue. *Hum Reprod*, Vol.23, No.11, pp: 2402-2414,

Wyns, C.; Curaba, M.; Vanabelle, B.; Van Langendonckt, A.&Donnez, J. (2010). Options for fertility preservation in prepubertal boys. *Hum Reprod Update*, Vol.16, No.3, pp: 312-328,

Wyns, C.; Curaba, M.; Martinez-Madrid, B.; Van Langendonckt, A.; Francois-Xavier, W.&Donnez, J. (2007). Spermatogonial survival after cryopreservation and short-term orthotopic immature human cryptorchid testicular tissue grafting to immunodeficient mice. *Hum Reprod*, Vol.22, No.6, pp: 1603-1611,

Zeng, W.; Snedaker, A. K.; Megee, S.; Rathi, R.; Chen, F.; Honaramooz, A.&Dobrinski, I. (2009). Preservation and transplantation of porcine testis tissue. *Reprod Fertil Dev*, Vol.21, No.3, pp: 489-497,

Permissions

The contributors of this book come from diverse backgrounds, making this book a truly international effort. This book will bring forth new frontiers with its revolutionizing research information and detailed analysis of the nascent developments around the world.

We would like to thank Igor I. Katkov, Ph.D., for lending his expertise to make the book truly unique. He has played a crucial role in the development of this book. Without his invaluable contribution this book wouldn't have been possible. He has made vital efforts to compile up to date information on the varied aspects of this subject to make this book a valuable addition to the collection of many professionals and students.

This book was conceptualized with the vision of imparting up-to-date information and advanced data in this field. To ensure the same, a matchless editorial board was set up. Every individual on the board went through rigorous rounds of assessment to prove their worth. After which they invested a large part of their time researching and compiling the most relevant data for our readers. Conferences and sessions were held from time to time between the editorial board and the contributing authors to present the data in the most comprehensible form. The editorial team has worked tirelessly to provide valuable and valid information to help people across the globe.

Every chapter published in this book has been scrutinized by our experts. Their significance has been extensively debated. The topics covered herein carry significant findings which will fuel the growth of the discipline. They may even be implemented as practical applications or may be referred to as a beginning point for another development. Chapters in this book were first published by InTech; hereby published with permission under the Creative Commons Attribution License or equivalent.

The editorial board has been involved in producing this book since its inception. They have spent rigorous hours researching and exploring the diverse topics which have resulted in the successful publishing of this book. They have passed on their knowledge of decades through this book. To expedite this challenging task, the publisher supported the team at every step. A small team of assistant editors was also appointed to further simplify the editing procedure and attain best results for the readers.

Our editorial team has been hand-picked from every corner of the world. Their multi-ethnicity adds dynamic inputs to the discussions which result in innovative outcomes. These outcomes are then further discussed with the researchers and contributors who give their valuable feedback and opinion regarding the same. The feedback is then collaborated with the researches and they are edited in a comprehensive manner to aid the understanding of the subject.

Apart from the editorial board, the designing team has also invested a significant amount of their time in understanding the subject and creating the most relevant covers. They scrutinized every image to scout for the most suitable representation of the subject and create an appropriate cover for the book.

The publishing team has been involved in this book since its early stages. They were actively engaged in every process, be it collecting the data, connecting with the contributors or procuring relevant information. The team has been an ardent support to the editorial, designing and production team. Their endless efforts to recruit the best for this project, has resulted in the accomplishment of this book. They are a veteran in the field of academics and their pool of knowledge is as vast as their experience in printing. Their expertise and guidance has proved useful at every step. Their uncompromising quality standards have made this book an exceptional effort. Their encouragement from time to time has been an inspiration for everyone.

The publisher and the editorial board hope that this book will prove to be a valuable piece of knowledge for researchers, students, practitioners and scholars across the globe.

List of Contributors

Luiz Augusto U. Santos
Institute of Orthopedics and Traumatology, Hospital das Clínicas of the School of Medicine of the University of Sao Paulo, dentist, Tissue Bank Technical Responsible and. Sao Paulo/SP, Brazil

Alberto T. Croci
Institute of Orthopedics and Traumatology, Hospital das Clínicas of the University of São Paulo school of Medicine, Professor and Tissue Bank Director - Sao Paulo/SP, Brazil

Nilson Roberto Armentano
School of Dentistry of the University of Santo Amaro- São Paulo, Brazil

Zeffer Gueno de Oliveira
Orthopedic Nurse Specialist, São Paulo/SP, Brazil

Arlete M.M. Giovani
Nurse, Institute of Orthopedics and Traumatology, Hospital das Clínicas of the University of São Paulo school of Medicine, Tissue Bank Coordinator - São Paulo/SP, Brazil

Ana Cristina Ferreira Bassit
Veterinarian, Tissue Bank Researcher, University of Florida, Gainesville, FL – Flórida-US

Thais Queiróz Santolin and Graziela Guidoni Maragni
Nurse, Institute of Orthopedics and Traumatology, Hospital das Clínicas of the University of São Paulo school of Medicine, Tissue Bank Team - São Paulo/SP, Brazil

Lucas da Silva C. Pereira
Dental Student, Institute of Orthopedics and Traumatology, Hospital das Clínicas of the University of São Paulo school of Medicine, Tissue Bank Team - São Paulo/SP, Brazil

Mohammad Hadi Bahadori
Cellular and Molecular Biology Research Center, Faculty of Medical Science, Guilan University of Medical Sciences, Rasht, Iran

Kelvin G.M. Brockbank
Cell & Tissue Systems, Inc., North Charleston, SC, USA
Georgia Tech / Emory Center for the Engineering of Living Tissues, The Parker H. Petit, Institute for Bioengineering and Bioscience, Georgia Institute of Technology, Atlanta, GA, USA
Medical University of South Carolina, Department of Regenerative Medicine and Cell Biology, Charleston, SC, USA

Lia H. Campbell
Cell & Tissue Systems, Inc., North Charleston, SC, USA

Alexandre C. Bitar
Vita Institute, Brazil

Rosa Maria Vercelino Alves, Fábio Gomes Teixeira, Paulo Henrique Kiyataka, Marisa Padula and Mary Ângela Fávaro Perez
Packaging Technology Center – Institute of Food Technology, Campinas, SP, Brazil

Monica Beatriz Mathor
Nuclear and Energy Research Institute - IPEN/CNEN-SP – São Paulo, Brazil

Cesar Augusto Martins Pereira
Biomechanics Laboratory - Institute of Orthopedics and Traumatology, Hospital das Clínicas of the University of São Paulo school of Medicine, Brazil

Renata Miranda Parca
National Health Surveillance Agency- ANVISA

Enrique Criado Scholz
CERAM: Centre for assisted reproduction in Marbella, Spain

Kampon Kaeoket
Faculty of Veterinary Science, Semen laboratory, Department of Clinical Science and Public Health, Mahidol University, Thailand

Thierry Joly, Vanessa Neto and Pascal Salvetti
Université de Lyon, France, VetAgro Sup – Isaralyon, UPSP ICE 'Interactions between Cells and their Environment', Team Cryobio, France

Hideaki Yamashiro
Laboratory of Animal Genetics and Reproduction, Faculty of Agriculture, Niigata University, Japan

Eimei Sato
Laboratory of Animal Reproduction, Graduate School of Agricultural Science, Tohoku University, Japan

Milan Milenkovic, Cesar Díaz-Garcia and Mats Brännström
Department of Obstetrics and Gynecology, Sahlgrenska Academy University of Gothenburg, Sweden

Vanessa Neto
Université de Lyon, VetAgro Sup – Veterinary Campus of Lyon, UPSP ICE 'Interactions between Cells and their Environment', Team Cryobio, France

Printed in the USA
CPSIA information can be obtained
at www.ICGtesting.com
JSHW011430221024
72173JS00004B/750

9 781632 394804